The CAN Bus Companion

Projects with Arduino Uno & Raspberry Pi

Dr Dogan Ibrahim
&
Ahmet Ibrahim, BSc., MSc.

elektor
design > share > earn

● This is an Elektor Publication. Elektor is the media brand of
Elektor International Media B.V.
PO Box 11, NL-6114-ZG Susteren, The Netherlands
Phone: +31 46 4389444

● **Declaration**

The authors and publisher have used their best efforts in ensuring the correctness of the information contained in this book. They do not assume, or hereby disclaim, any liability to any party for any loss or damage caused by errors or omissions in this book, whether such errors or omissions result from negligence, accident or any other cause.

All the programs given in the book are Copyright of the Author and Elektor International Media. These programs may only be used for educational purposes. Written permission from the Author or Elektor must be obtained before any of these programs can be used for commercial purposes.

● British Library Cataloguing in Publication Data
A catalogue record for this book is available from the British Library

● **ISBN 9783895765414** Print
ISBN 9783895765421 eBook

Editor: Jan Buiting, MA
Prepress Production: D-Vision, Julian van den Berg

Contents

Preface

The Controller Area Network (acronym: CAN) structure was originally developed for use in passenger cars. Today, sophisticated CAN controller chips are available from over 20 manufacturers, and CAN is finding applications in other fields such as medical, aerospace, process control, automation, and so on. With the establishment of CAN within the Automation (CiA) association back in 1992, manufacturers and users rallied to exchange ideas and develop the CAN standards and specifications.

This book is about the use of the Arduino Uno development board and Raspberry Pi 4 with CAN bus interface modules for the design of practical CAN bus based projects. Examples of popular hardware and software development kits are described concisely. Using these kits simplifies the embedded design cycle considerably and considerably eases developing, debugging, and testing a project based on, or relying on, the CAN bus. Projects are given using both the 2- and 3-node CAN bus variants.

This book is written for students, practicing engineers, enthusiasts, and for everyone else who may need to learn more about the CAN bus and its applications. The book assumes that the reader has some knowledge of basic electronics. Knowledge of the C programming language and Python and familiarity with the Arduino Uno and Raspberry Pi will be an advantage, especially if the reader intends to develop microcontroller-based projects using the CAN bus.

The book should be a useful source and reference for anyone interested in finding an answer to one or more of the following questions:

- What bus systems are available for the automotive industry?
- What are the principles of the CAN bus?
- What types of frames (or data packets) are available in a CAN bus system?
- How can errors be detected in a CAN bus system, and how reliable is a CAN bus system?
- What types of CAN bus controllers exist?
- What are the operational principles of the MCP2515 CAN bus controller?
- How do I create a CAN bus project using Arduino Uno?
- How do I create 2-node and 3-node Arduino CAN bus projects?
- How do I analyze the data on the CAN bus?
- How do I create a CAN bus project using Raspberry Pi?
- How do I create 2-node and 3-node Raspberry Pi CAN bus projects?

We hope that you will find the book helpful and enjoyable and will be able to create your next CAN bus project using Arduino Uno and/or Raspberry Pi with CAN bus interface modules.

Dogan Ibrahim & Ahmet Ibrahim
London, 2022

Chapter 1 • Automotive Bus Systems

1.1 Overview

Today's vehicles are complex machines incorporating mechanical and electronic parts. The number of electronic components used in vehicles has rapidly increased in recent years. As a result of the increase in safety, comfort, and performance requirements, we see many more electronic components being added to modern vehicles. Henceforth, there has been an increasing demand to connect these electronic components in such a way that they can communicate with each other reliably, safely, and in real time.

In the past, electronic units used to be connected in a complex way, with hundreds of wires running to distinct parts of a vehicle. Consequently, it was difficult to trace an electronic fault. There was no coordination between various parts of the electronics, as each electronic part was controlled independently of the others. Maintenance and repair of the vehicle electronics were extremely difficult, as in many cases it was not easy to locate and change a faulty component.

Passenger safety has also become one of the most important considerations in today's vehicles. In the past decade, safety equipment has evolved and moved from physical to electronic assisted safety, starting with braking and type technology, through collision protection and airbags, and most recently, on to safety-related driver assistance systems. The latest modern vehicles are smart and intelligent machines and are equipped with many sensors that can evaluate the surroundings and display useful and safety-related information to drivers. These sensors form intelligent local networks together with actuators, displays, and fast digital processors, such as high-speed purpose-built microcontrollers.

Figure 1.1 shows a traditional old-style vehicle electronic system with sensors and actuators interconnected to each other in a complex manner. One of the major problems in this type of design is maintenance. The wiring was so complex that it was nearly impossible to trace and rectify faults.

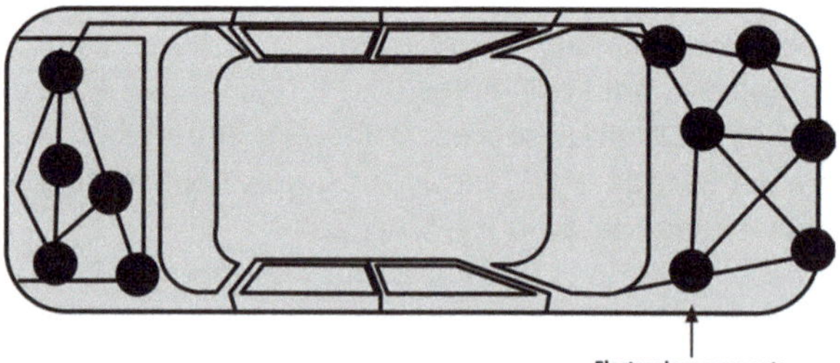

Figure 1.1: Traditional old-style vehicle electronics.

As the complexity of vehicle electronics has increased, it has become difficult for manufacturers to design safe and reliable electronic systems based on old traditional methods. Current requirements cannot be obtained using a simple electronic control unit. The solution is to interconnect the various electronic modules with a high-performance network. This is why it has become necessary to design a network-based electronic system where electronic modules can easily be attached to a network and then controlled from a central, intelligent unit (e.g., Engine Control Unit). The result of this is an "intelligent" car where many sensors and actuators are used to sense the environment and perform multiple functions. An example is the automatic turning on of headlights when it becomes dark or when the car goes through a tunnel. Another example is the automatic operation of windscreen wipers when the rain begins, and so on.

One of the advantages of a network-based vehicle electronic system is that it is relatively easy to trace and detect a faulty module. In addition, the wiring is a lot simpler and easier to maintain. For example, by communicating with the central intelligent controller unit, one can tell whether or not the overall electronics system is healthy, and if not, the faulty modules can easily be detected. A networked system also allows the various modules on the bus to communicate with each other and exchange information if required. For example, the intelligence controller unit can receive the engine temperature value from the temperature sensor module. This temperature can then be displayed on an electronic dashboard. Should the temperature be too high, appropriate messages can be sent to the responsible parts of the engine and corrective measures can easily be taken. Figure 1.2 shows a modern vehicle where an electronic network system is used to interconnect and control the electronic modules.

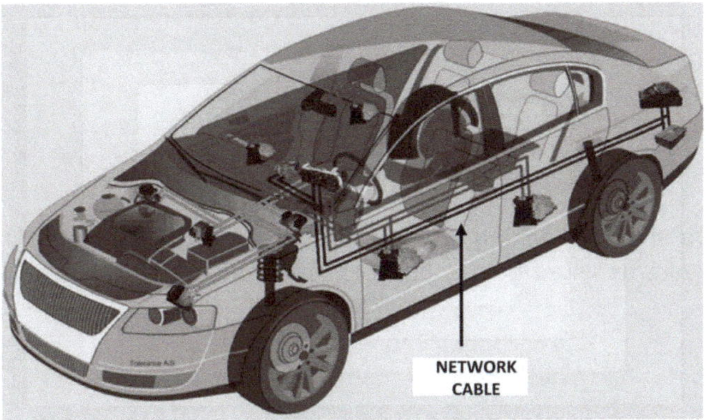

Figure 1.2: Modern vehicle with an electronic network.

This chapter provides an overview of the most important vehicle network systems currently used in vehicles. Moreover, it provides a table to compare the advantages and disadvantages of each system.

1.2 Vehicle network systems

Vehicle network systems (or networks) were classified in 1994 by the Society of Automotive Engineers (SAE). According to this classification, bus systems were classified based on their bandwidth (i.e., data rate) and functions of the network system. The classification divides bus networks into four: Class A, Class B, Class C, and Class D.

Class A networks are low-speed, low-cost networks with data rates of less than 10 kbps. These systems are mainly used in the body of the car.

Class B networks operate between 10 and 125 kbps and are used for information exchange, e.g., instrument cluster, vehicle speed, and so on.

Class C networks operate between 125 and 1 Mbps and are used in a wide range of applications, such as engine control.

Class D networks operate above 1 Mbps, and they are mainly used in telematics applications.

There are many automotive network systems (or bus systems, since the vehicle networks are in the form of bus wires), some developed by vehicle manufacturers on their own, and some developed jointly with semiconductor manufacturers. Popular vehicle network systems include:

- CAN bus
- LIN bus
- FlexRay
- MOST
- Byteflight
- DSI bus
- Intellibus
- SAE J1850
- BST bus
- NML bus
- And others...

In this section, we shall be looking at the brief details of the most commonly used automotive network systems listed below, and concentrate on the CAN bus, which is the topic of this book.

- LIN bus
- FlexRay
- MOST
- Byteflight
- Intellibus
- CAN bus

1.2.1 LIN bus

The Local Interconnect or "LIN" bus is a low-cost bus operating at 20 kbps that has been developed to create a standard for low-cost, low-end communication in automotive electronics. LIN fits in the low end of the automotive network systems, thus making it cost-effective. This bus is mainly used for non-important body/comfort functions. LIN is a single-wire, single-master/multiple-slave type bus system where the vehicle chassis is used as the return path. In a typical application, the master broadcasts a message with a message header, asking for data, and the slave that has the correct message header sends the requested data including a checksum for error checking. Typical LIN bus applications include the control of small motors for wipers, sunroof control, heating control, rain sensor control, steering wheel, seats, doors, etc. where high bandwidth is not required. The LIN bus is used in applications where the implementation of CAN bus would be too expensive. The initial LIN specification was defined by a consortium consisting of BMW, Audi, Volvo, VW, Motorola, Volcano, and DaimlerChrysler. The LIN bus is based on the Serial Communications Interface (UART) with 8-bit data.

1.2.2 FlexRay

As the demand for improved safety, increased performance, and enhanced comfort rises, then the need for more complex and more sophisticated electronic vehicle network systems arises. FlexRay was initially developed by BMW and Daimler Chrysler in 1999 as a fast, efficient, and error-free automotive bus system. FlexRay is suited to real-time, high-speed applications as it supports a bandwidth of up to 10 Mbps. Both electrical and optical transmission mediums can be used. FlexRay supports single and dual-channel configurations, with the dual channel offering enhanced fault-tolerance and increased bandwidth. Considered to be the next step beyond LIN and CAN, it is mainly used in safety-critical applications and in real-time high-speed engine control. Some application areas of FlexRay are: engine control, ABS, transmission control, break control, suspension control, etc.

1.2.3 MOST

The Media Oriented Systems Transport (MOST) bus is mainly used in automotive telemetric and multimedia applications, such as audio control, video, navigation, communication, and so on. Modern vehicle systems include multimedia devices such as DVD/CD players, TVs, navigation systems, graphics displays, and mobile computing which require very high data transfer rates. This is achieved using the MOST network. The original MOST network was developed by BMW and DaimlerChrysler in 1998.

MOST supports a maximum of 64 devices on the bus with very high bandwidth. Typical data rates are 28.8 Mbps synchronous, and 14.4 Mbps asynchronous. Up to 50 Mbps and even 150 Mbps are currently under development. An optical medium is used for data transmission that is free of any electromagnetic radiation or interference.

1.2.4 Byteflight

Byteflight was developed by BMW and offers 10 Mbps bandwidth. Byteflight is mainly used in safety-related networks, such as automotive and avionic systems (e.g., in vehicle airbags, body electronics, and so on). For it to be safe, the data protocol must be fault-tolerant and deterministic. Data control mechanisms before Byteflight had been either event-controlled or time-controlled. Event-controlled (e.g., CAN) only transmits data when data is ready or when a data request arrives. Time-controlled data protocols grant time to each node in accordance with a pre-defined sequence. The number of messages to be transmitted cannot generally be changed during operation, since the number of allocated time slots is fixed. The Byteflight protocol combines both event-controlled and time-controlled protocols and guarantees deterministic latencies for a specific number of high-priority messages, and flexible use of bandwidth for lower-priority messages.

The first use of the Byteflight in mass production was by BMW in their Series-7 cars and within a networked passive safety application. Byteflight supports various network protocols in a mixed bus environment.

1.2.5 Intellibus

Intellibus is a high-speed multi-drop communications bus offering up to 15 Mbps bandwidth. It was initially conceived by Boeing to reduce the wiring complexity associated with distributed systems in aerospace applications. It is a low-cost bus that allows a large number of sensors to be connected, and it is used mainly in engine control, transmission, and other parts requiring high speed. A typical Intellibus network consists of a Network Interface Controller (NIC) and 1 to 255 Intellibus Interface Modules (IBIM) that can be connected to sensors. The NIC can be installed on a PC or some other electronic device. Normally, the NIC is downloaded with software to sample data from various IBIMs in a scheduled manner, at specific intervals. In a complex network, two or more NIC cards can be used to increase node capacity. Special dedicated software is available to program the NIC and IBIMs. The maximum bus length is 30 m (at 12.5 Mbps and with 64 nodes). The Intellibus is used in automotive electronics, process control, automation, avionics, medical fields, and several other fields.

1.2.6 CAN bus

The Controller Area Network (CAN) bus is the main subject of this book. In this section, we examine the basic properties of this bus. Detailed descriptions of the CAN bus and CAN bus-related projects are provided in the remaining chapters of the book.

CAN is a serial, two-wire multi-master bus developed by Robert Bosch GmbH in the 1980s. It is one of the most widely used automotive electronic system buses today. The physical layer of CAN consists of a pair of twisted cables. CAN provides dependable, robust, and fast communication up to 1 Mbps (with a 40 m bus length). CAN 2.0A is the original CAN version consisting of the fields:

- Start of Frame bit
- 18-bits header (having 11-bit message identifier)
- 0-8 bytes data
- 15-bit Cyclic Redundancy Check (CRC)
- 3-bit acknowledgement slot
- 7-bit of End of Frame

The CAN bus is based on the CSMA/CR (Carrier Sense Multiple Access/Collision Resolution) "mechanism" to prevent frame collisions during transmissions on the bus. Each CAN node monitors the bus and when the node detects that the bus is idle, it may start transmitting data. If other nodes on the bus attempt to send data at the same time, arbitration will take place and the node with the highest priority (i.e., lowest message identifier) will win the arbitration and send its own data. The CAN bus has a simple error detection-and-recovery mechanism. Receiving nodes check the integrity of the messages by looking at the CRC fields. If an error is detected, the other nodes on the bus are informed through error flag messages. Figure 1.3 shows a typical CAN bus implementation with two nodes, A and B. CAN is a Class A/B type network.

Summary of CAN bus features:

- Multi-master bus where each node can be a master
- Multicast reception
- Speeds up to 1 Mbits/s
- Highly dependable

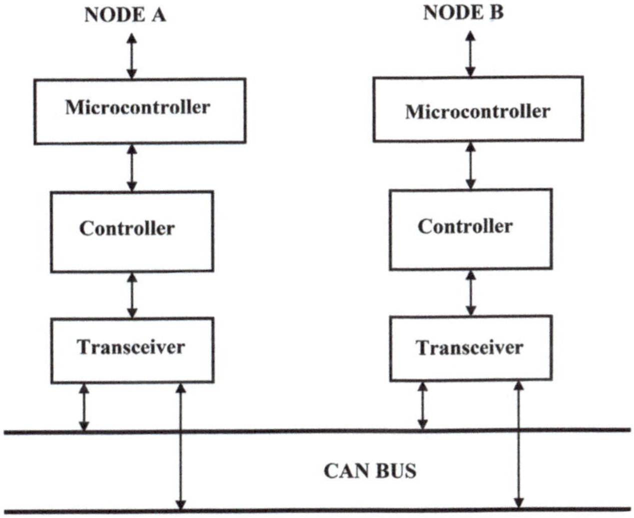

Figure 1.3: CAN bus with two nodes.

1.2.7 Others

Other automotive bus standards that should be mentioned in this chapter are the SAE J1850 (or simply J1850) which was developed in 1994, and the MI bus (Motorola Interconnect). J1850 was widely used in cars such as GM, Chrysler, and Ford, and this bus is used for diagnostics and data-sharing applications. There are two versions of this standard: PWM (Pulse Width Modulation) with 41.6 Kbps using a two-wire differential physical layer, and VPW (Variable Pulse Width) with 10.4 Kbps using a single-wire physical layer. The two standards are incompatible with each other. The J1850 protocol is frame-based and uses CSMA/CR arbitration where a frame consists of a Start of Frame, a header byte, data bytes, one byte CRC, and End of Data symbol (a 200-µs Low pulse). Most OBD (On Board Diagnostic) tools support the J1850 protocol for diagnostic purposes. The J1850 standard is old and is being phased out. The J1850 standard is a Class B type network.

The MI bus is a single-wire bus with one master and many slaves. This is a low-cost and low-data rate bus mainly used to drive seats, mirrors, etc. The master is the controller which sends addresses and data to all slaves on the bus. Slaves with matching addresses respond to the request.

1.3 Comparison of automotive bus systems

Currently, one of the most commonly used bus systems is the CAN bus, which can be used at speeds up to 1 Mbps.

Table 1.1 is a summary of commonly used automotive networks (or bus systems), comparing the various systems. For each system, the Class, General information, Bandwidth, and typical Application areas are given.

In general, the LIN bus is used in low-speed automation, such as wiper motor control, rain sensors, etc. CAN is used in engine control, clutch control, and so on. FlexRay is used in very high-performance and safety-critical applications. This is a relatively new bus structure and is hoped to be the automotive standard of the future. MOST is generally used in automotive multimedia, and navigation applications. This bus is optical-based and is tolerant to electrical noise. Byteflight is also a new automotive network system and is well suited to high-demand real-time safety critical applications, such as airbag control. Byteflight is based on fiber optics and as such is much more tolerant to electromagnetic radiation and electrical noise. Intellibus, finally, offers high bandwidth and is mainly used in aerospace applications. It is not as safe and dependable as Byteflight.

	Class	General	Bandwidth	Application
CAN	A, B	- Twisted pair - Low-cost - Widely used - Multiple masters	1 Mbps	- Body - Engine
LIN	A	- Low-cost - Low-speed - Master/slave - Single wire	10 kbps	- Body - Comfort
FlexRay	D	- Very high-speed - Future standard - Master/slave (for synchronization)	10 Mbps	- Engine - Safety
MOST	D	- Data efficient - Standard for multimedia - Master/slave (for synchronization) - Fiber-optic	Mbps	Multimedia
Byteflight	D	- High-speed - Master/slave - Fiber-optic	10 Mbps	Safety critical
Intellibus	D	- High-speed	15 Mbps	- Body - Engine - Safety

Table 1.1: Automotive bus systems

Figure 1.4 and Figure 1.5 show examples of passenger cars where several bus systems are used (source: *Introduction to CAN – Application Note REJ05B0804-0100/Rev. 1.00*). In Figure 1.4, the CAN bus can be seen at the left of the figure, connected to the instruments, climate control, lights, steering wheel, power train, and the lock system. The MOST bus can be seen at the lower-left part of the figure, connected to multimedia equipment such as speakers, digital radio, vehicle computer, and navigation equipment. In this particular application, FlexRay can be seen at the lower-right of the figure, connected to the engine, steering system, and brakes. A diagnostic tool, connected to the CAN bus at the bottom-right of the picture, is used to check the state of the vehicle.

In Figure 1.5, headlights, air-conditioning, screen wipers, doors, and windows are controlled by the LIN bus, while FlexRay controls the brakes. Most of the other parts of the car are controlled by the CAN bus.

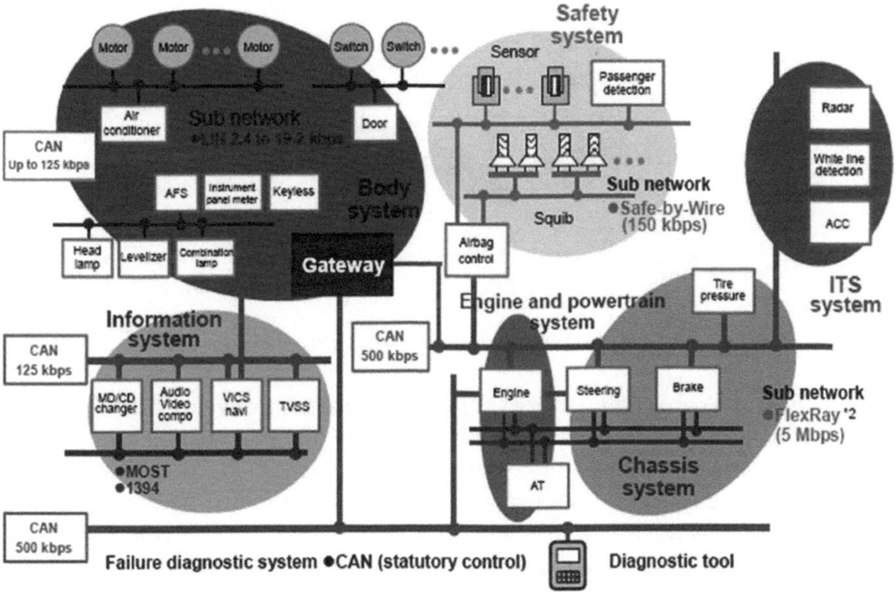

Figure 1.4: A typical passenger car with several bus systems.

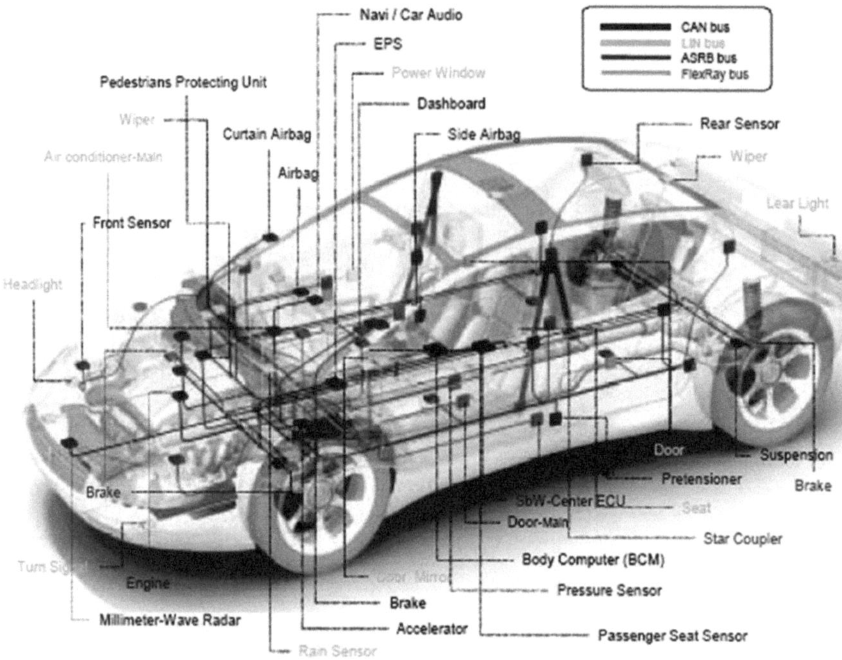

Figure 1.5: Another passenger car with several bus systems.

Today, more than 20 semiconductor manufacturers produce devices with CAN interfaces. Almost all new passenger cars manufactured in Europe and the Far East are equipped with at least one CAN network. CAN is one of the most dominant bus protocols used in passenger cars. Since the year 2000, CAN usage in vehicles has rapidly increased from approximately 120 million CAN devices to around 800 million devices in the year 2007. Currently, this figure is over tens of billions of devices.

CAN and CAN-based higher-level protocols such as CANopen and DeviceNet are becoming more widely accepted by North American manufacturers and it is expected that these protocols will have very high penetrations in most of the industrial and commercial automation markets. In particular, CAN will be used in the following diverse applications within the next five to ten years:

- Passenger cars
- Buses
- Trains
- Maritime electronics
- Aviation electronics
- Factory automation
- Lifts
- Medical equipment
- Programmable machine controllers
- Home entertainment systems

- Domestic appliances
- Military applications
- Space applications

An interesting and important application of the CAN bus will almost certainly be in the medical industry. Medical equipment and devices such as X-ray machines, ultrasound, radiotherapy machines, CAT scanners, MRI machines, and so on could be designed and developed to make use of the CAN bus for their internal communication structures. This approach will simplify the design of machines as well as increase their safety and reliability, also helping to improve ease of maintenance.

1.4 The basic structure of a CAN bus automotive system

It is worthwhile to take a look at the basic structure of a CAN bus-based automotive system before going into the details and theory of the CAN protocol and electronics.

With the development of the CAN bus, there have been major advances in both reliability and safety of automotive electronics. As shown in Figure 1.6, in a CAN bus based automobile, all units are connected to each other over a two-wire bus system. As you will discover in later chapters in more detail, the two wires of the bus are named CAN_LO and CAN_HI. There is full coordination between all the units, and the ECU (Engine Control Unit) is responsible for controlling each unit, making sure that all units operate as expected. In Figure 1.6, the units at the upper part of the figure are time-critical, requiring high priority. Similarly, the units at the lower part of the figure are slow-speed non-critical units requiring lower priorities.

The complete wiring and state of the vehicle can be monitored and interrogated by connecting a diagnostic tool through the diagnostic connector located on the engine. This tool can be a laptop running special software developed by the manufacturer, or a dedicated hand-held diagnostic device can be used to communicate directly with the ECU. Using this tool and with the help of the ECU, it is possible to get information about the state of each unit of the vehicle. For example, we can easily interrogate the Airbag Module and check that it is working properly.

CAN is a two-wire bus system (see Figure 1.7) with resistive connectors at each end of the bus. All the units (or nodes, or CAN-stations) are connected rigidly to the bus using CAN connectors (e.g., T-connectors, End-connectors, and so on).

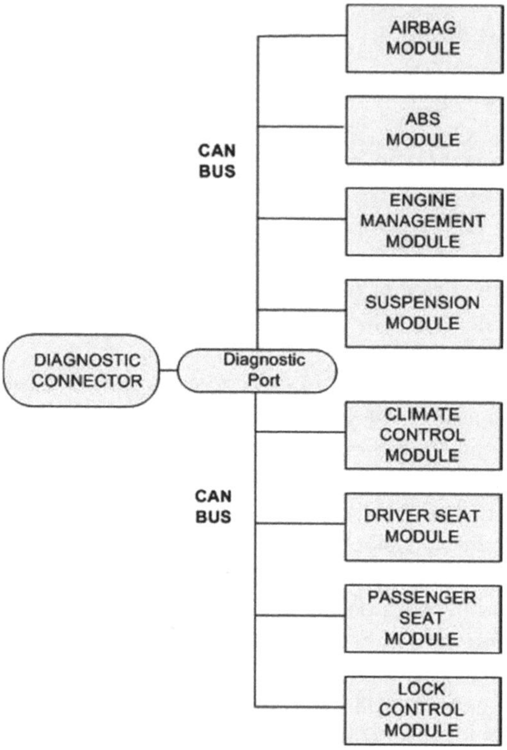

Figure 1.6: CAN bus-based automotive wiring system.

In general, information is received from various sensors (e.g., engine temperature sensor) located in various parts of the vehicle. As shown in Figure 1.8, the sensor information is then passed to a microcontroller, and then to the CAN bus via a CAN controller and transceiver. The function of the CAN transceiver is to make physical connection to the actual CAN bus for communication over the bus. The CAN controller is under the control of the microcontroller and performs the CAN protocol specific functions.

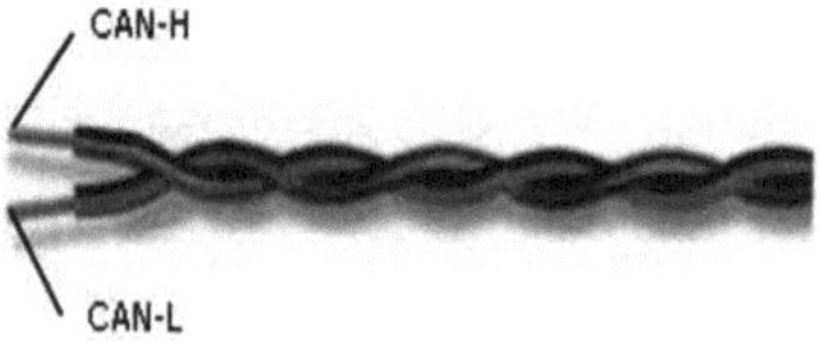

Figure 1.7: CAN is a two-wire bus.

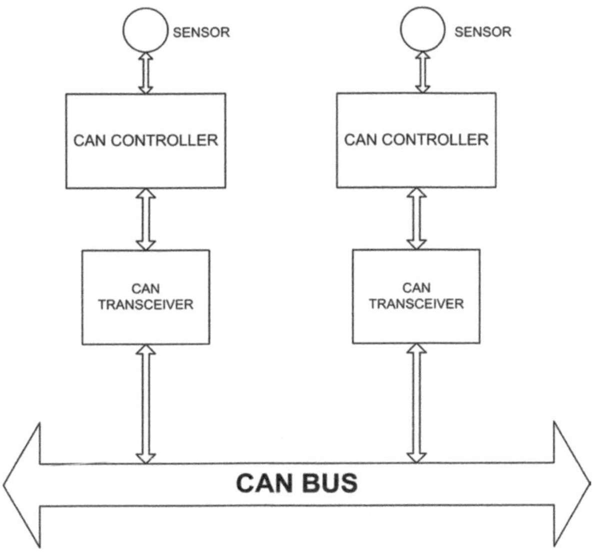

Figure 1.8: Sensors on the CAN bus.

A typical ECU is shown in Figure 1.9. This ECU is suitable for car engines with up to 6 cylinders (ECU model: ECU MS3 Sport, Source: *Bosch Motorsport, Equipment For High Performance Vehicles, Edition 2011/11*).

Figure 1.10 shows diagrammatically how the ECU (in the center of the figure) can be connected to other parts of a car. The port labeled "Diagnosis" is used for OBD (On Board Diagnostics) where a hand-held terminal (a scan tool) or a laptop computer can be connected using a special connector and the overall "health" of the vehicle can easily be examined.

Figure 1.9: ECU MS3 Sport.

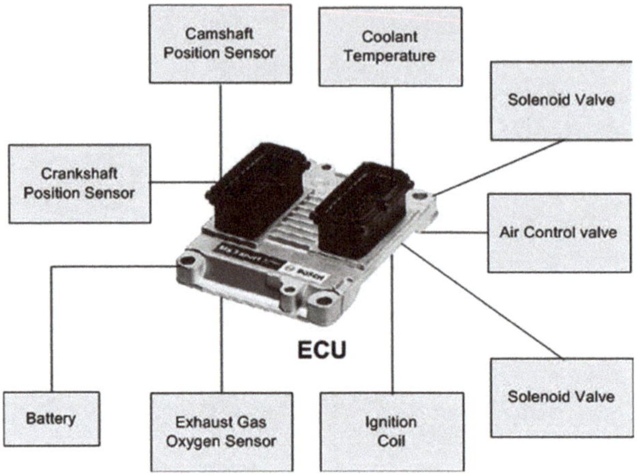

Figure 1.10: ECU and other parts of a car.

1.5 CAN bus advantages

The main benefits of the CAN bus can be summarized as follows:

- Low wiring complexity
- Easy to manage twisted-pair wire bus
- New nodes can be easily added and removed
- Breakdown of a node does not affect the other nodes in the system
- Centralized control
- All devices on the bus can read the same message
- Fail-safe against electromagnetic radiation
- Deals effectively with errors
- Easy diagnostic and maintenance

1.6 CAN bus disadvantages

Some of the disadvantages of the CAN bus include:

- the data rate can be low in some applications (limited to 1 Mbps);
- expensive to implement as special controllers are required;
- complete system shutdown if the main CAN trunk wire is damaged.

1.7 CAN bus main properties

The CAN bus is currently used in many diverse fields, from passenger cars to marine electronics, medical electronics, aviation electronics, and so on. The reason for the wide acceptance of the CAN bus is arguably its high performance, besides reliability, robustness, and safety. The main properties of the CAN bus can be summarized below (the details of these properties will be examined in the next chapters):

- Two-wire twisted-cable bus
- Bus-terminated with resistors at each end

- Up to 1 Mbps speed on the bus (for a 40-m length bus). Higher data rates for shorter bus lengths
- Transfer of up to 8 data bytes at any time
- Messages sent and received as frames under a clearly defined robust protocol
- Multi-master priority-based access to the bus
- Any one node can transmit while other nodes listen
- Bus arbitration if two or more nodes attempt to transmit at the same time
- Broadcast-type message transfer where all nodes can receive the same message
- No node addresses. Nodes accept or reject data on the bus based on message acceptance filtering
- Remote data request where a node can request data from another node
- Error detection and signaling
- Automatic retransmission of messages if failed to transmit (because of the bus arbitration)
- Automatic deactivation of nodes that present consistent errors

1.8 CANopen

CANopen is a high-level communication protocol and specification based on the earlier CAN protocol. CANopen was developed for the vehicle industry in order to simplify and standardize the communication between various devices developed by different manufacturers. Although originally developed for the vehicle industry, it has been used in a wide range of other industries.

Some of the features of CANopen are:

- Up to 128 nodes in the same CAN network
- Multi-master architecture
- Bus length up to 5 km (reduced bit rate at longer lengths)
- Easy high-level access to all devices
- Mixing devices from different manufacturers (manufacturer independence)
- Device synchronization
- High reliability

Although CANopen provides network management and manufacturer independence to a CAN network, it also has the disadvantage that bandwidth is reduced.

Chapter 2 • CAN Physical Layer Structure

2.1 Overview

The physical layer forms the hardware interface that establishes a connection between a number of nodes. A minimum of two nodes are required to communicate on a bus.

Figure 2.1 shows a CAN bus with three nodes. The bus is made up of a twisted-pair cable and is terminated at both ends with a resistor so that the characteristic resistance of the bus works out at 120 ohms. The two wires of the bus are termed CAN_H and CAN_L.

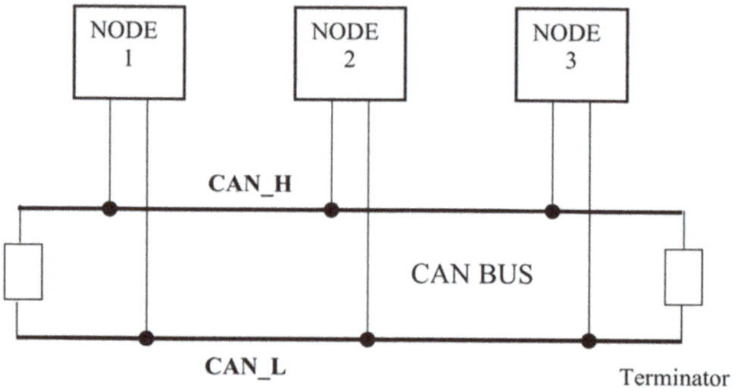

Figure 2.1: CAN bus with three nodes.

2.2 CAN bus termination

Resistive termination is required to minimize signal reflections on the bus wires. Usually, a single 120-ohm resistor is connected to each end of the bus. Although the power rating of the chosen resistor is not particularly important, possible short circuits to power supplies on the bus should be considered when selecting a resistor power rating. A 0.25-watt, 5% tolerance resistor is generally used, but it is recommended to use a higher rating (e.g., 1 watt) resistor to avoid any damage to the bus due to possible transceiver short circuits.

Although a single 120-ohm resistor is usually used, in general, one of the following methods can be used to terminate the bus:

- Standard termination
- Split termination
- Biased split termination

The most commonly used termination method is the standard termination where a 120-ohm resistor is used at each end of the bus as shown in Figure 2.2.

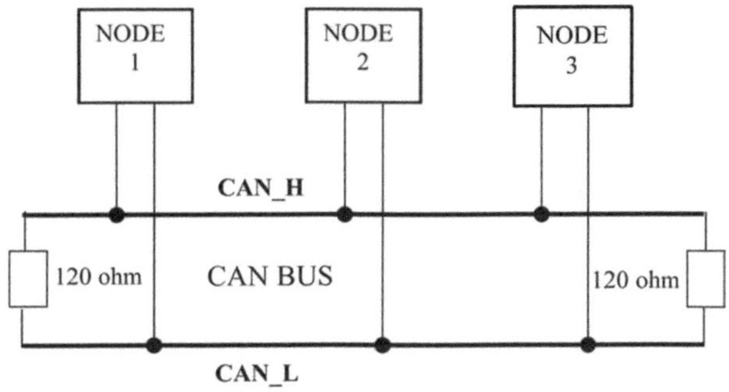

Figure 2.2: Standard bus termination.

Figure 2.3 shows split termination that is gaining in popularity. In this method, two 60-ohm resistors and a capacitor are used at each end of the bus. The advantage of this method is that it eliminates high-frequency noise from bus lines. Care must be taken to match the resistors so as not to reduce the effective immunity on the bus. Typically, a 4.7-nF capacitor is chosen which generates a response with a 3 dB point at approximately 1.1 Mbps.

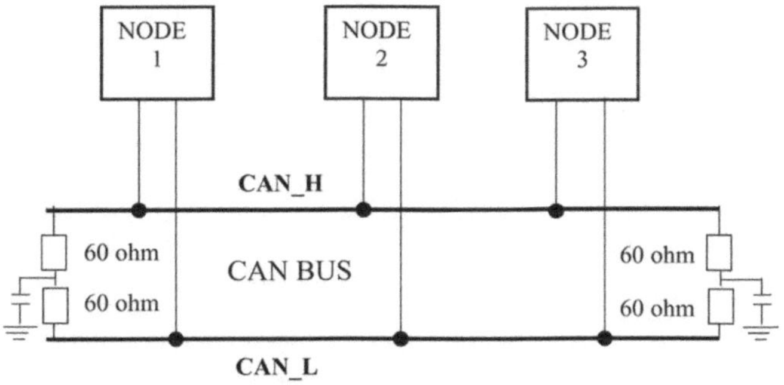

Figure 2.3: Split bus termination.

Figure 2.4 shows the biased split bus termination, where a voltage divider circuit and a capacitor are used at each end of the bus. As in split termination, the biased split termination increases the EMC performance of the bus.

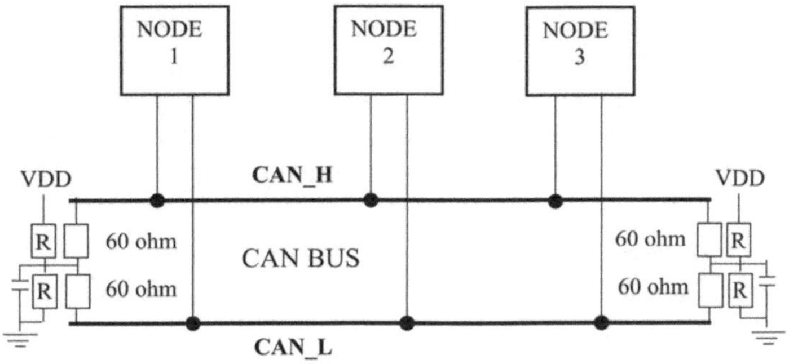

Figure 2.4: Biased split bus termination.

It is important to note that terminating resistors must be placed at the farthest ends of the wires. In Figure 2.5, one of the terminating resistors is at the wrong place. A wrong terminating resistor value in the wrong termination place is likely to cause errors on the bus.

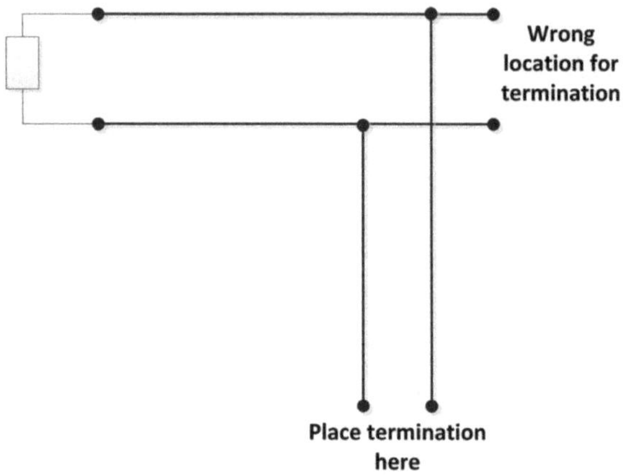

Figure 2.5: Wrong placement of a terminating resistor.

Correct termination of the bus can be checked by using a digital voltmeter to measure the bus resistance. Set the meter to measure resistance and connect the two leads of the meter to the ends of a node on the bus (see Figure 2.6). The meter should measure 60 ohms.

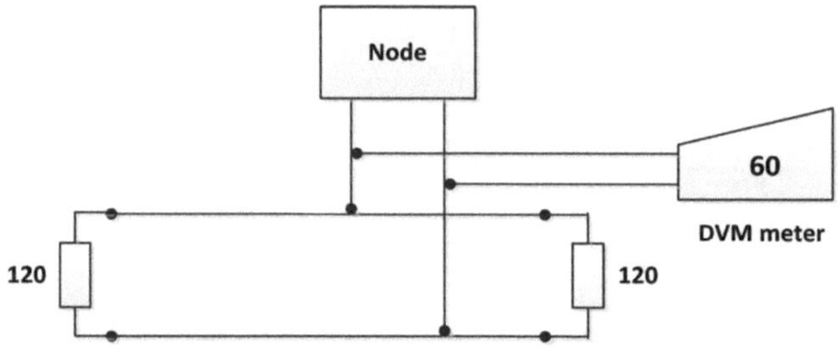

Figure 2.6: Checking the bus termination.

Although unshielded cables are used in most CAN bus applications, there are many harsh environments that may require rugged interconnection. Shielded cables can be used in such environments, and it is recommended that the shield of the cable be connected to the ground at one point of the shield, located as close to the cable as possible (see Figure 2.7). This prevents parasitic currents such as ground loops from being formed in the shield and consequently current to flow in the shield between the two grounds.

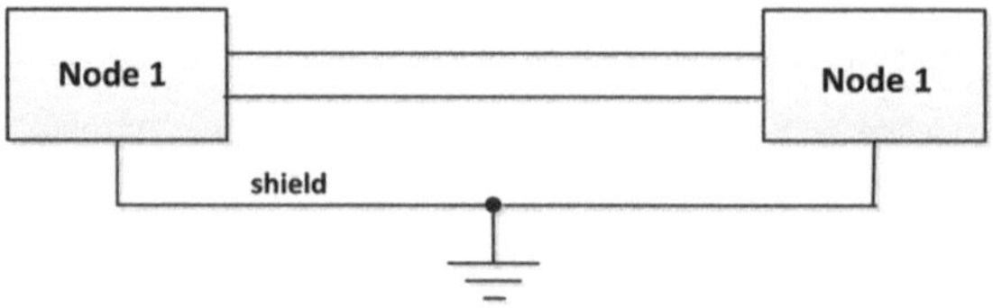

Figure 2.7: CAN bus correct grounding.

2.3 CAN bus data rate

ISO-11898 CAN specifies that a device on the bus must be able to drive a 40-m cable length and achieve 1 Mb/s data rate. In practice, much longer lengths are achieved by lowering the bus speed. Table 2.1 shows bus speed against bus length and a nominal bit time. At 1000 kbps (1 µs nominal bit time) the maximum allowed bus length is 40 m, whereas at 10 kbps (100 µs nominal bit time) the maximum allowable bus length rises to 6700 m.

Data rate (kbps)	Nominal bit time (µs)	Bus length (meters)
10	100	6700
20	50	3300
50	20	1300
125	8	530
250	4	270
500	2	130
1000	1	49

Table 2.1: Data rate against nominal bit time and maximum bus length.

A graph of the maximum data rate against the maximum allowed bus length is shown in Figure 2.8.

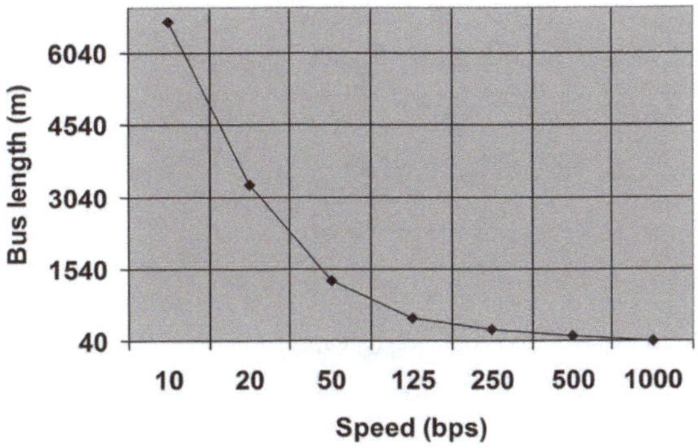

Figure 2.8: CAN bus data rate and maximum bus length.

In long cables, the bus lines should be as close as possible to a straight line to keep possible reflections to a minimum. The cable length can be extended using a bridge device or repeater.

2.4 Cable stub length

In high data rate applications, the length of cable stubs and the distance between the nodes becomes an important factor. Since the stub lines are not terminated, signal reflections can develop on these lines, creating bus errors. At the maximum data rate of 1 Mbps, the length of the cable stubs (see Figure 2.9) should not be greater than 0.3 m, and the maximum node distance should be 40 m.

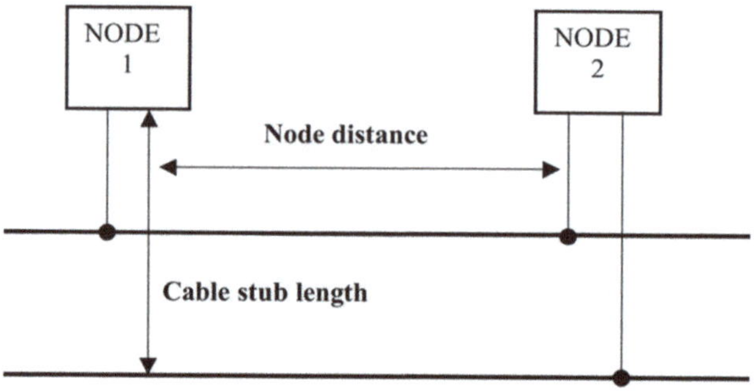

Figure 2.9: Cable stubs and node distances.

2.5 CAN bus node

A CAN bus node consists of a CAN controller and a CAN transceiver (Figure 2.10). The CAN controller is connected to a microcontroller and the CAN transceiver is connected to the physical CAN bus. Some microcontrollers have built-in CAN controller circuitry (e.g. ARM), and thus a CAN node can be created using an external CAN transceiver chip. If the micro-controller does not have built-in CAN controller circuitry, then an external CAN controller and transceiver chips will be required to create a CAN node. CAN modules are available with built-in controller and transceiver chips. Such modules usually communicate with the microcontroller using the SPI bus.

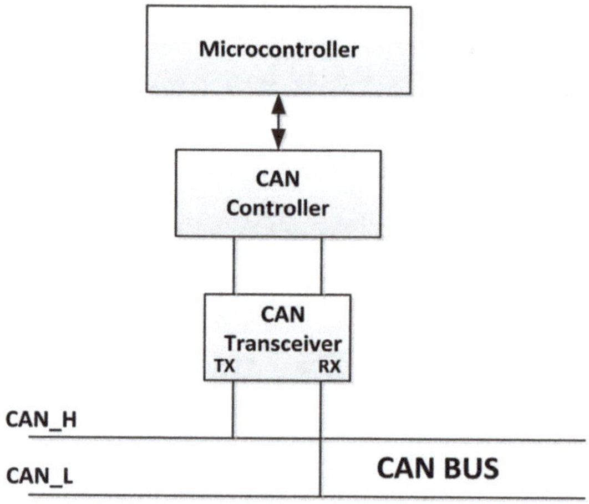

Figure 2.10: CAN bus node.

2.6 CAN bus signal levels

The data on the CAN bus is differential and the bus specifies two logical states: dominant and recessive. Figure 2.11 shows the state of signals on the bus (see document ISO-11898-4 for more details).

The recessive state is logic "1" and at this state, the differential voltage on the bus (i.e., V_{diff} = CAN_ H – CAN_L) is around 0 V (ideally CAN_H = CAN_L = 2.5 V). In practice, the recessive differential output voltage is less than 0.05 V at a bus transmitter output device.

The dominant state is logic "0" and at this state, the differential voltage on the bus (i.e., V_{diff} = CAN_ H – CAN_L) is around 2 V (ideally CAN_H = 3.5 V and CAN_L = 1.5 V). In practice, the dominant differential output voltage is between 1.5 V and 3.0 V.

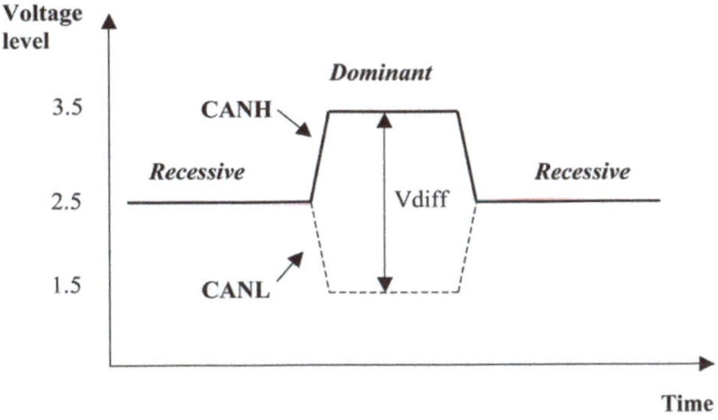

Figure 2.11: CAN bus signal levels.

2.6.1 CAN_H voltage

CAN_H voltage should be half of the 5 V during times when no message is transmitted (or when the CAN bus is in recessive state). This voltage can be measured using a digital voltmeter (DVM) if it is already known that the CAN controller is not sending any messages. In order to measure CAN_H voltage, disconnect the TX pin from the transceiver and connect a +5 V supply to the transceiver TX input to force the transceiver to the recessive state (see Figure 2.12). Connect one lead of the meter to the CAN_H pin of the transceiver, and the other pin of the meter to the ground pin of the transceiver. The measured voltage should be around 2.5 V.

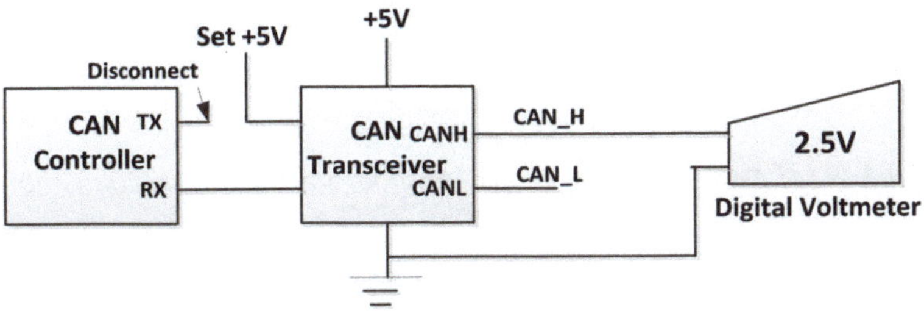

Figure 2.12: Measuring the CAN_H voltage.

2.6.2 CAN_L voltage

CAN_L voltage should be half of the 5 V during the times when no message is transmitted (or when the CAN bus is in recessive state). As above, this voltage can be measured using a digital voltmeter. Use the same procedure as above, but this time, connect the voltmeter between CAN_L and the ground terminals (see Figure 2.13). The measured voltage should be around 2.5 V.

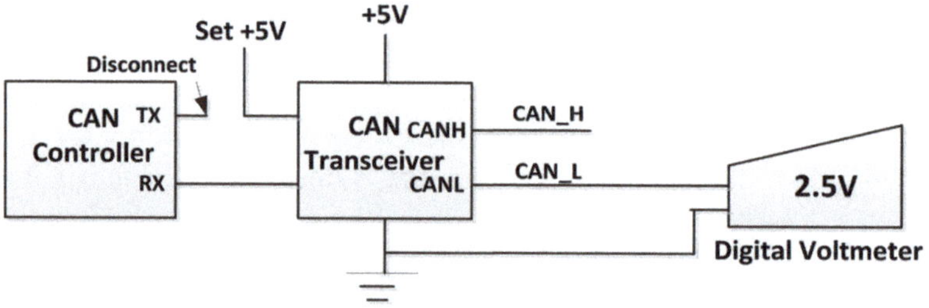

Figure 2.13: Measuring the CAN_L voltage.

2.6.3 CAN signal waveform

CAN signal waveforms can be displayed on an oscilloscope by connecting CAN_H and the CAN_L bus lines to a suitable oscilloscope. Figure 2.14 shows a typical display during message transfer on the bus. Notice that a dominant bit moves the voltage from 2.5 V to 3.5 V and a recessive bit moves the voltage from 2.5 V to 1.5 V.

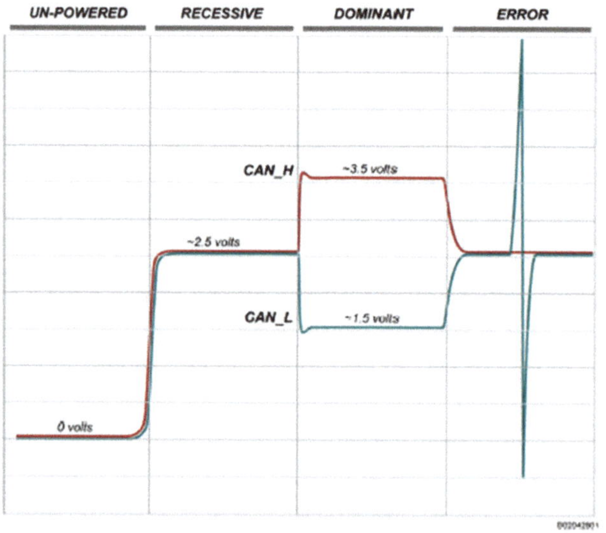

Figure 2.14: CAN signal waveform. Source: Vector Informatik Gmbh).

2.6.4 Bus arbitration

When several bus nodes attempt to transmit at the same time, bus arbitration logic is used to grant access of a device to the bus. When there is arbitration on the bus, a dominant bit state always wins out over a recessive bit state.

2.6.5 Bus transceiver

The output of a CAN transceiver circuit is usually in open-collector (e.g., TTL logic) or in open-drain (e.g., CMOS logic) format. When several such devices are connected to a bus, the net logic state of the bus is defined by the logical "AND" of the device outputs (also called the "Wired AND"). For example, if three devices are connected to the bus, the state of the bus will be logic "1" if (and only if) all the outputs of all the three devices are at logic "1", otherwise the bus will be at logic "0".

Figure 2.15 shows the output stage of a typical CAN transceiver. Source: MCP2551, Microchip Technology Inc.).

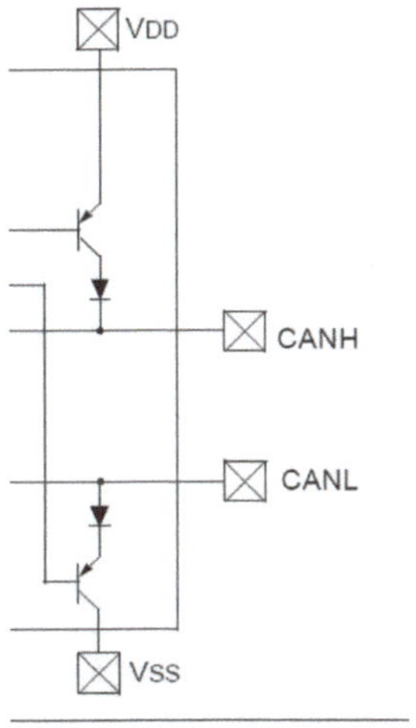

Figure 2.15: Output stage of a typical CAN bus transceiver.

2.7 CAN connectors

Even though CAN is a two-wire network, a power signal and a ground signal are often added to the standard CAN connectors. As mentioned earlier, the actual wires used for the bus can either be unshielded twisted-pair (UTP) or shielded twisted-pair (STP). Shielded cables should be used in electrically noisy environments and when long bus cables are needed.

The standard CAN connector is a 9-pin D-type connector (DE-9) as shown in Figure 2.16. The pin configuration is shown in Table 2.2. Pin 2 and pin 7 are the CAN_L and CAN_H signals, respectively. Pin 3 and pin 9 are used as signal ground and signal power, respectively. Signal ground, and signal power pins can be useful when it is required to power remote devices. Care should be exercised to make sure that the current capacity of the cable used is not exceeded.

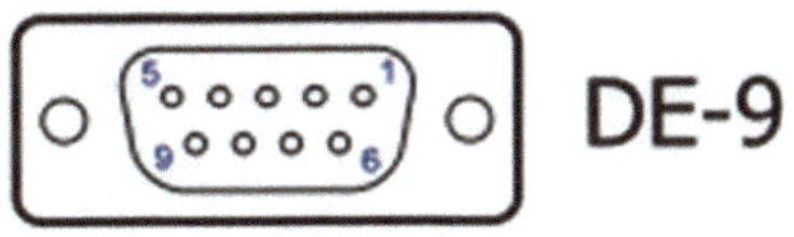

Figure 2.16: CAN D-connector.

9 Pin (male) D-Sub CAN bus PinOut		
Pin #	**Signal Names**	**Signal Description**
1	Reserved	Upgrade Path
2	CAN_L	Dominant Low
3	CAN_GND	Ground
4	Reserved	Upgrade Path
5	CAN_SHLD	Shield, Optional
6	GND	Ground, Optional
7	CAN_H	Dominant High
8	Reserved	Upgrade Path
9	CAN_V+	Power, Optional

Table 2.2: CAN-D connector pin configuration.

Figure 2.17 shows a CAN-D type connector with a short CAN cable and terminating resistors.

Figure 2.17: CAN-D type connector with CAN bus cable.

Some companies use a 10-pin header to link devices to the bus. Table 2.3 gives the standard pin configuration for this type of connector.

10-Pin Header CAN bus PinOut		
Pin #	**Signal Names**	**Signal Description**
1	Reserved	Upgrade Path
2	GND	Ground, Optional
3	CAN_L	Dominant Low
4	CAN_H	Dominant High
5	CAN_GND	Ground
6	Reserved	Upgrade Path
7	Reserved	Upgrade Path
8	CAN_V+	Power, Optional
6	Reserved	Upgrade Path
7	Reserved	Upgrade Path

Table 2.3: CAN 10-pin header pin configuration.

It is also common to use either RJ10 or RJ45-type connectors in CAN bus applications, as shown in Figure 2.18. The pin configuration of this type of connector is shown in Table 2.4.

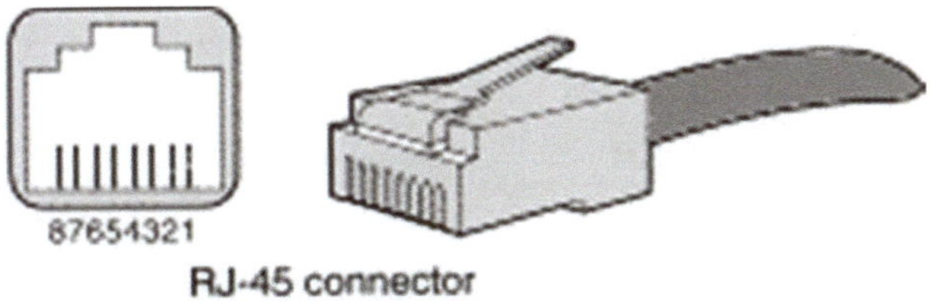

Figure 2.18: RJ10 or RJ45 type CAN connector.

RJ10, RJ45 CAN bus PinOut			
RJ45 Pin #	**RJ10 Pin #**	**Signal Name**	**Signal Description**
1	2	CAN_H	Dominant High
2	3	CAN_L	Dominant Low
3	4	CAN_GND	Ground
4	-	Reserved	Upgrade Path
5	-	Reserved	Upgrade Path
6	-	CAN_SHLD	CAN Shield, Optional
7	-	CAN_GND	Ground
8	1	CAN_V+	Power, Optional

Table 2.4: RJ10 or RJ45 connector pin configuration.

Some companies prefer to use a 5-pin circular connector, as shown in Figure 2.19.

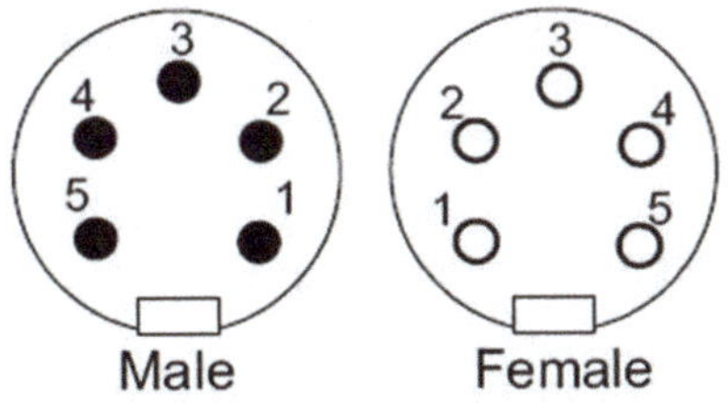

Figure 2.19: 5-pin circular connector.

Table 2.5 shows the pin configuration of 5-pin circular connectors.

5-Pin Micro/Mini CAN bus PinOut		
Pin #	Signal Names	Signal Description
1	CAN_SHLD	Shield, Optional
2	CAN_V+	Power, Optional
3	CAN_GND	Ground
4	CAN_H	Dominant High
5	CAN_L	Dominant Low

Table 2.5: 5-pin circular connector pin configuration.

Professional quality CAN cables, connectors, and terminating resistors are commercially available from many manufacturers and electronic product distributors. An example commercial CAN network is shown in Figure 2.20 from ESD Electronic System Design Gmbh (https://esd.eu/en/products/can-cable-accessories).

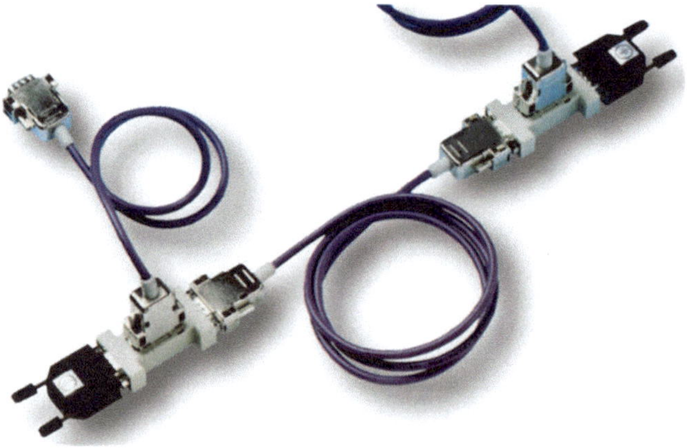

Figure 2.20: Commercial CAN network.

2.8 CAN repeaters

CAN repeaters allow several CAN networks to be connected to each other. A typical repeater is shown in Figure 2.21, and has the following features:

- Linking 2 or 3 CAN networks
- With or without electrical isolation
- Bit rate up to 1 Mbits/s
- Wide supply voltage (8 V to 30 V)

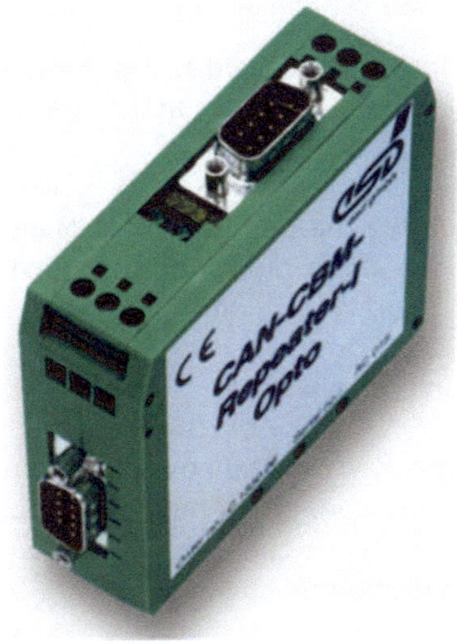

Figure 2.21: CAN repeater (https://esd.eu/en/products/can-repeater).

2.9 CAN PC interface

Some companies provide plug-in CAN PC cards for PCIe slots on a PC. These cards include CAN bus controllers and transceivers and enable a PC to be used as a CAN bus node. Some cards include up to 4 controllers. An example card (CPC-PCIe) is shown in Figure 2.22.

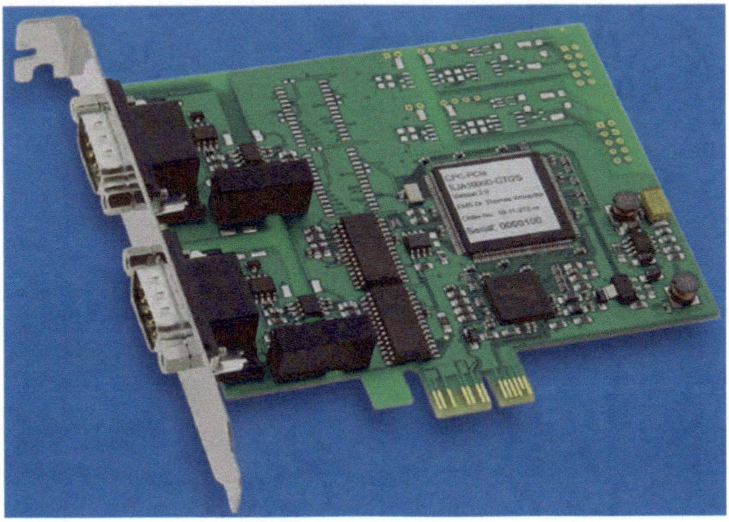

Figure 2.22: CPC-PCIe card. (http://www.ems-wuensche.com).

Chapter 3 • CAN Bus Frames

Messages on CAN bus are sent and received using **frames**. A frame is basically like a **packet** in a TCP/IP-type network, where the actual data is encapsulated with control data.

CAN bus communication is not like the popular Client-Master type of communication. In the CAN bus, all nodes have the same rights, and they can transmit as well as receive data at suitable times.

Some important CAN protocol features are:

- CAN bus is multi-master. When the bus is free, any device attached to the bus can start sending a message. When multiple devices attempt to send data at the same time, then collisions can occur on the bus. Collisions are detected and avoided using an arbitration mechanism.

- The CAN bus protocol is flexible. Devices connected to the bus have no addresses (or node IDs), which means that messages are not transmitted from one node to another based on addresses. Instead, all nodes on the bus receive every message transmitted on the bus, and it is up to each node to decide whether or not the received message should be kept or discarded. A single message can be destined for a particular device on a particular node, or for many nodes, depending on how the bus system is designed. Messages have **message identifiers**, and **acceptance filters** on each node decide whether or not to accept a message being transmitted on the bus. Another advantage of having no addresses is that when a device is added to or removed from the bus, no configuration data needs to be changed (i.e., the bus is "hot pluggable").

- Messages sent on the bus have priorities. A message with a lower message identifier has a higher priority.

- CAN bus communication speed is not fixed. Any communication speed up to the allowed maximum can be set for the devices attached to the bus.

- CAN bus offers Remote Transmit Request (RTR), which means that a node on the bus is able to request data from other nodes. Thus, instead of waiting for a node to continuously send data, a request for data can be sent to the node. For example, in a vehicle, where the engine temperature is an important parameter, the system can be designed so that the temperature is sent periodically over the bus. However, a more elegant solution is to request the temperature as needed. This second approach will minimize bus traffic and CPU loading thus increasing performance, while maintaining the integrity.

- All devices on the bus can detect an error. The device that has detected an error immediately notifies all other devices. Nodes that transmit faulty data, or nodes that are always receiving data in error will remove themselves from the bus, thus allowing the normal bus operations to continue.
- Receiving nodes on the bus check the validity of the received frame and acknowledge the consistency. The transmitting node monitors the bus during the acknowledgement slot.
- Multiple devices can be connected to the bus at the same time, and there are no logical limits to the number of devices that can be connected. In practice, the number of nodes that can be attached to a bus is limited by the delay time of the bus and electrical load on the bus.

There are four message frames in CAN bus:

- **Data Frame**: Defines the data transfer between nodes.
- **Remote Frame**: Used by a node to request transmission of a message (i.e., data) from another node.
- **Error Frame**: Any node on the bus can send an error frame to signal an error condition.
- **Overload Frame**: This frame is used by a receiving node to indicate that it is not yet ready to receive frames.

Let's look at each frame in greater detail.

There are essentially two types of CAN protocols: 2.0A and 2.0B. CAN 2.0A is the earlier standard with 11 bits of identifier (see next section), while CAN 2.0B is the new extended protocol with 29 bits of identifier. 2.0B controllers are completely backwards compatible with 2.0A controllers and can receive/transmit messages in either format.

We shall firstly look at the standard CAN 2.0A frames and then explore the CAN 2.0B frames in later sections.

There are two types of 2.0A controllers. The first is capable of sending and receiving 2.0A messages only, and the reception of a 2.0B message will flag an error. The second type of 2.0A controller (known as 2.0B passive) sends and receives 2.0A messages but will also acknowledge receipt of 2.0B messages and then ignore them.

3.1 Data Frame

The data frame is used by a transmitting device to send data to receiving devices on the bus, and this frame is the most important frame managed by the user. The data frame can be sent in response to a request, or it can be sent whenever it is required to send the value of some parameter to other nodes on the bus (e.g., the temperature can be sent at periodic intervals).

Figure 3.1 shows the structure of a data frame. The bus is normally idle. Then, a standard data frame starts with the *Start of Frame* (SOF) bit, which is followed by an 11-bit *identifier* and *Remote Transmission Request* (RTR) bit. The **control** field is 6-bits wide and indicates how many bytes of data are in the data field. The **data** field can be 0 to 8 bytes, and it contains the actual data to be sent. The data field is followed by the 16-bit checksum *(CRC)* field, which checks whether or not the received bit sequence is corrupted. The **ACK** field is 2-bits wide and is used by the transmitting node to receive acknowledgement of a valid frame from any receiver. The end of the message is indicated by a 7-bit **End of Frame** (EOF) field. Successive frames must be separated by at least 3-bit times, called the **inter-frame space (ITM)**.

The total number of bits required by the data frame are (assuming successive frames are to be sent):

SOF	1 bit
Identifier	11 bits
RTR	1 bit
Control	6 bits
Data	0 to 64 bits (0 to 8 bytes)
CRC	16 bits
ACK	2 bits
EOF	7 bits
ITM	3 bits

In total, 47 bits (no data) to 111 bits (8 bytes of data) are required by the data frame.

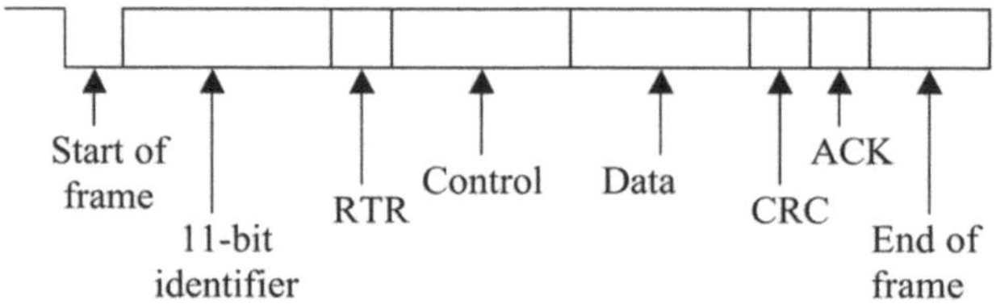

Figure 3.1: Standard data frame.

Figure 3.2 shows the standard data frame in more detail.

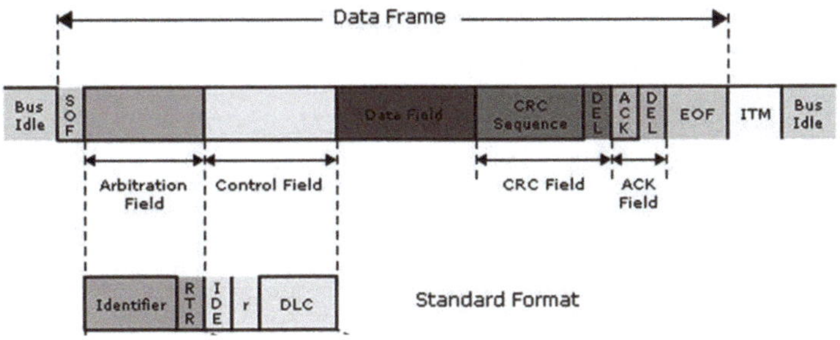

Figure 3.2: Detailed standard data frame.

3.1.1 Start of Frame (SOF)

The Start of Frame field is 1 bit and indicates the beginning of a data frame sent while the bus is in the idle state. The bus is said to be in the idle state if a sequence of 11 recessive bits is present on the bus (comprising 1-bit ACK delimiter, 7-bits EOF, and 3-bits ITM). The SOF bit is in the dominant state.

As we shall see later, the SOF field also starts the arbitration sequence on the bus when multiple devices attempt to send data at the same time.

3.1.2 Arbitration Field

The arbitration field of a standard data frame is 12-bit wide, and it consists of the following two components:

- 11-bit Identifier
- 1-bit Remote Transmission Request (RTR)

The 11-bit identifier (MSB sent first) is used to identify messages on the bus. Different devices can send messages with different identifiers. For example, a temperature sensor device can send a message with an identifier of 20, while a pressure sensor can send a message with an identifier of 25. The receiving nodes have **Acceptance Filters**, and by programming these filters we can accept or reject messages with given identifier numbers. For example, if we program the acceptance filters of two nodes to accept messages with identifier numbers of 20, then whenever the temperature data is sent by the above temperature sensor node, our two nodes will accept and use this temperature data. With the 11-bit identifier, up to 2048 (2032 useable) unique identifier numbers can be set.

A data frame with a lower identifier has a higher message priority and, as we shall see shortly, such a message is granted bus access by the arbitration mechanism.

Arbitration is used to resolve bus conflicts that occur when several devices at once start sending messages on the bus. During the arbitration phase, each transmitting device transmits its identifier and compares it with the existing level on the bus. If the levels are equal, the device continues to transmit its identifier. If the device detects a dominant level on the bus while it is trying to transmit to a recessive level, it quits transmitting and becomes a receiving device. After arbitration, only one transmitter is left on the bus. This transmitter continues to send the remainder of its frame bits. The process of arbitration is illustrated in Figure 3.3 by an example consisting of three nodes with the following identifiers and with RTR = 0:

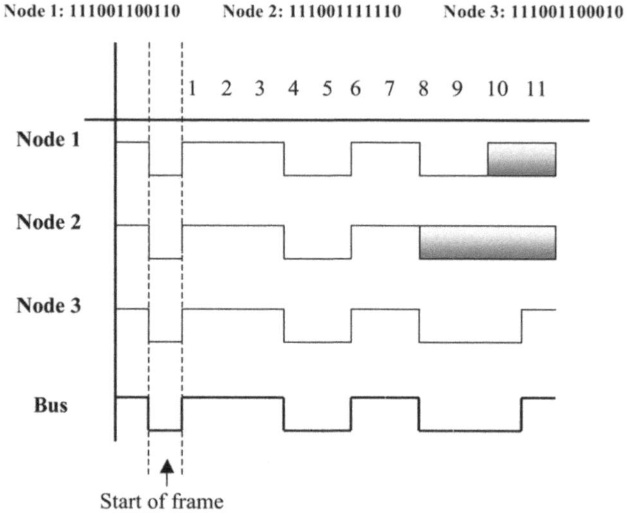

Figure 3.3: Example CAN bus arbitration.

Recalling that the recessive level corresponds to 1 and the dominant level to 0, and that the dominant level has higher precedence than the recessive level, the arbitration in Figure 3.3 is performed as follows:

- Bits are transmitted starting from the MSB. Numbering of the MSB bit as bit 1:
- All nodes start transmitting simultaneously, first sending their SOF bits.
- Then they start sending their identifier bits. Up to the 8th bit, all nodes send the same identifier bit. The 8th bit of Node 2 is in recessive state, while the corresponding bits of Nodes 1 and 3 are in dominant state. Therefore, Node 2 stops transmitting and returns to receive mode. Receiving mode is indicated in the figure by a gray color field.
- The 9th bits of Nodes 1 and 3 are the same, so they continue sending their identifier bits.
- The 10th bit of Node 1 is in the recessive state, while the same bit of Node 3 is in dominant state. Thus, Node 1 stops sending and returns to receive mode (gray color field).
- The bus is now left to Node 3, which can send the remainder of its identifier, and the remainder of the frame bits (RTR bit, control field, data field, CRC, and the ACK bits).

Looking at the **identifiers** of each node again with their hexadecimal equivalents, you have:

Node1: 11100110011	**Node2: 1110011111**	**Node3: 11100110001**
Hex: 733	**73F**	**731**

The node with the smallest identifier number is Node 3, and thus this node has the highest priority. This is compatible with the given example from above, where only Node 3 is given transmit right on the bus.

Note that the nodes on the bus had no addresses in the above example. Instead, all the receiving nodes (Nodes 1 and 2 above) pick up all the data sent by Node 3, and they use their acceptance filters to determine whether or not they want to accept the data transmitted by Node 3.

It is important to note that, after Node 3 sends its data, the situation may become different, and for example, Node 2 can become the bus master by sending a data frame with a lower identifier. This is the beauty of the CAN bus, allowing a multi-master operation of the bus, enabling each node to become either a master or a client whenever they wish.

3.1.3 RTR Field

The 1-bit RTR field indicates the transmission of a data frame (RTR = 0), or a Remote Request Frame (RTR = 1). This field, together with the identifier bits, form what is known as the arbitration field. In the example above, RTR = 0.

3.1.4 Control Field

The control field is 6-bit wide, and as shown in Figure 3.4, consists of the IDE (Identifier Extension) bit, a reserved bit (r0), and 3 DLC (Data Length Code) bits.

	IDE	r0	DLC3	DLC2	DLC1	DLC0	

Figure 3.4: CAN control field bits.

The IDE bit indicates the CAN format used. IDE = 0 for the standard CAN 2.0A format, and IDE = 1 for the extended CAN 2.0B format.

r0 is a reserved bit.

DLC3 – DLC0 determine the number of bytes in the data field (0 to 8 bytes). Table 3.1 shows how bits DLC3 – DCL0 are used to indicate the size of data field. For example, if 4 bytes are to be sent in standard CAN 2.0A format then the control field has the following bits:

	0	0	0	1	0	0	

3.1.5 Data Field

The data field contains the actual data of the message. The data size can vary from 0 to 8 bytes. The data is transmitted with the MSB byte first.

3.1.6 CRC Field

The CRC (Cyclic Redundancy Check) field consists of the following bits:

- 15-bit CRC sequence
- 1-bit CRC delimiter

The CRC field is used to check the frame for a possible transmission error. The CRC calculation includes the SOF bit, arbitration field, control field, and data fields (see Figure 3.5). The receiving node calculates the CRC based on the received data and compares it with the CRC sent with the frame. If the two CRCs do not match, an error is assumed.

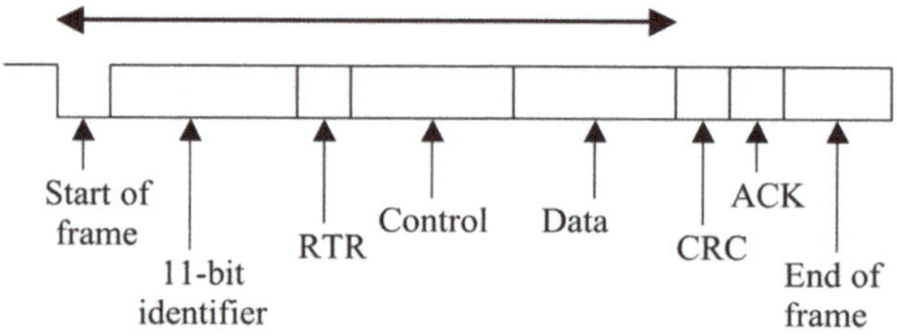

Figure 3.5: Fields used in the CRC calculation.

The receiving node calculates the CRC in the same way as the transmitting node. There are many CRC generator polynomials, but the one used by the CAN bus is as follows:

The message is regarded as a polynomial and is divided modulo-2 by the CRC generator polynomial. The CRC is generated by hardware and uses the following polynomial:

$$x^{15}+x^{14}+x^{10}+x^{8}+x^{7}+x^{4}+x^{3}+1$$

i.e., 16-bit; bit pattern: 1100010110011001

- The remainder of this modulo-2 division is the CRC sequence, which is transmitted together with the message.

- The receiver divides the message using the same generator polynomial.

- If a CRC error is detected, the receiver discards the message and transmits an Error Frame to request re-transmission.

A delimiter bit is sent at the end of the 15-bit CRC field. The delimiter is always recessive (i.e., 1). The reason for sending the delimiter bit is to allow more time for the CRC to be calculated.

3.1.7 ACK Field

The ACK field consists of the following bits:

- 1-bit ACK
- 1-bit delimiter

The ACK bit (sometimes called the ACK slot) is a confirmation that the frame has been received normally with no errors. i.e., it is confirmation that the CRC check is successful by the receiving nodes.

The transmitting node sends the ACK bit in recessive (i.e., 1) mode. During the ACK slot, the transmitting node switches to receive mode by sending a recessive signal to the bus. All the receiving nodes which have received a data frame with a **correct** CRC report this to the transmitter by sending a dominant bit during the ACK time slot. If the transmitting node detects a positive acknowledge, that is a dominant ACK, the transmitting node knows that at least one receiving node has got the message correctly.

When there are several receiving nodes, the various possibilities are as follows:

- If the transmitting node detects a dominant bit during the ACK slot, the transmitting node knows that at least one receiving node has got the message correctly (remember the bus "Wired-AND" logic). In fact, this means that all the receiving nodes received the message correctly. This is because if one node receives a CRC error, it will send an error frame, which will destroy the faulty message for every node (i.e., every node, including the transmitting node will receive the error frame and discard the message). In this case, the transmitting node will re-transmit the message after a specified timeout.

- If the transmitting node detects a recessive bit during the ACK slot, the transmitting node knows that all the receiving nodes have reported an error. In this case, it is probable that the transmitting node calculated the CRC wrongly, or there are no receivers, or there was data corruption on the bus. Therefore, the transmitting node will re-transmit the message after a specified timeout.

The 1-bit ACK delimiter is always in recessive state.

3.1.8 End of Frame Field

This is a 7-bit field consisting of 7 recessive bits. The data frame is always terminated by this field. When several data frames are to be sent sequentially, it is important that a 3-bit interframe gap is left between each frame. The interframe gap is always in the recessive state.

3.2 Remote Frame

The remote frame is used by a node to request the transmission of a message from another node. Remote frames are seldom used, and their use is not recommended by CiA (CAN in Automation).

The remote frame is identical to the data frame except for the following differences:

- The remote frame does not have a data field (see Figure 3.6). i.e., it consists of the SOF, identifier field, RTR, control field, CRC field, ACK field, and the EOF field.
- The RTR bit of the remote frame is set to recessive (i.e., 1) state.

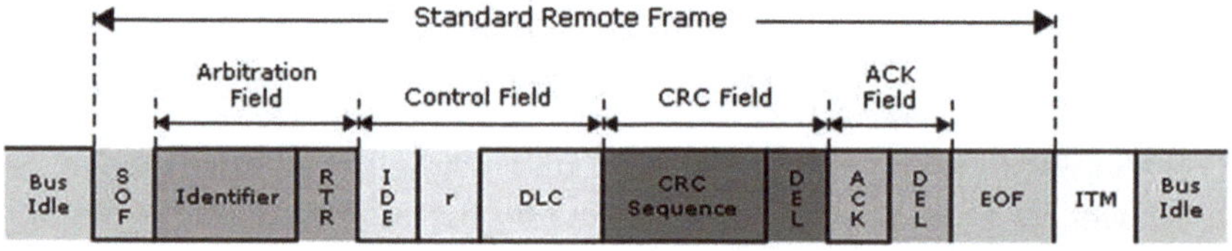

Figure 3.6: The remote frame.

Nodes receiving remote frames and accepting them will send a data frame with the requested data, but only one node can send the data.

The data request actually happens after sending two frames. The first frame is the remote frame (with RTR = 1) telling the other nodes that data is being requested. The second frame is a data frame with the same identifier, telling other nodes what the requested data is. The DLC of the remote frame must be the same as the DLC of the data frame sent just after the remote frame. If a remote frame and data frame with the same identifier attempt to access the bus at the same time, the data frame will be given access as its RTR is set to 0 and hence has a higher priority (remember that the RTR bit is part of the bus arbitration).

3.3 Error Frame

Error frames are generated and transmitted by the CAN hardware and are used to indicate when an error has occurred during transmission. The generation of an error frame terminates the faulty data on the bus and all the nodes on the bus discard the current data.

As shown in Figure 3.7, an error frame consists of a 6-bit Error Flag and an 8-bit Error Delimiter. A 3-bit interframe gap should also be used at the end of the error frame.

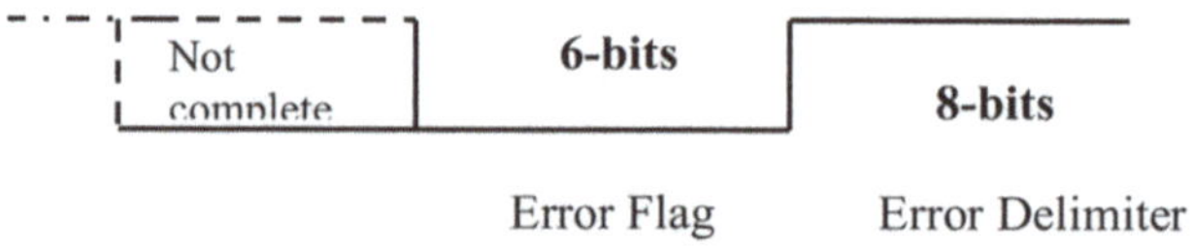

Figure 3.7: CAN Error frame.

According to the CAN standard, any 5 consecutive bits of the same polarity (5 consecutive dominant bits, or 5 consecutive recessive bits) are a violation of the bus standard and are considered an error condition. When the CAN standard is violated, the data currently on the bus is considered to be faulty and is discarded by all the nodes. After the interframe gap, the transmitting node re-transmits the same frame. In the error frame, 6 dominant bits are sent consecutively and as this violates CAN standards, the complete frame is discarded (as we shall see later, it is permissible to send more than 5 bits with the same polarity as actual data by using a method called "bit stuffing").

There are two types of error flags: active error flags, and passive error flags. An active error flag consists of 6 dominant bits, and a passive error flag consists of 6 recessive bits.

Error frames are sent immediately after an error has been detected. In the case of CRC type errors, the error frame is sent after the 2-bit time of detecting the error. This is so that there is no conflict with the ACK field.

It is important to realize that when an error frame is sent by a node, all other nodes on the bus recognize this error condition, and they too send error frames. It is possible that some of these nodes may also start sending error frames at the same time, but some nodes may start sending error frames after detecting the error condition on the bus, i.e., after 6 bits. Since these nodes will also start sending their error frames, it is possible that the error flag can be extended to a maximum of 12 bits on the bus. Figure 3.8 shows the case where the error frame can be extended to 12 dominant bits of error flag, and 8 recessive bits of error delimiter.

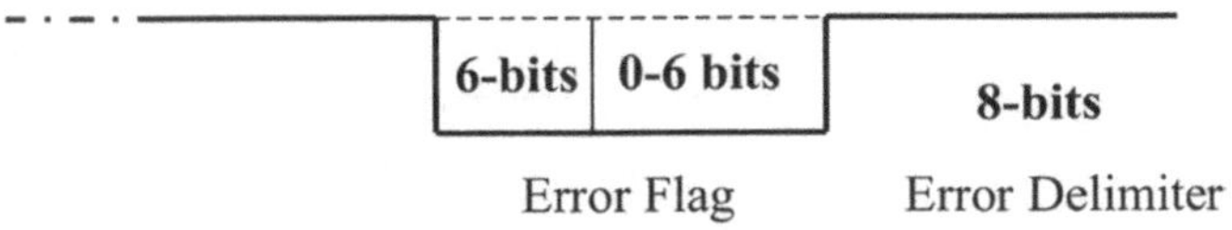

Figure 3.8: Extended error frame.

The error delimiter field consists of 8 recessive bits and is sent after the error flag. In the case when the error flag on the bus is more than 6-bits, all the nodes on the bus will send the first bits of their error delimiter bits and keep sending recessive bits until the bus turns into the recessive state (the bus will become recessive when all the nodes sent the first bit of their error delimiter bits). When the bus is in the recessive state, all the nodes will send the remaining 7 bits of their error delimiter bits.

It is important to realize that when the worst case is considered, the error frame consists of 20 bits (12 error flag bits + 8 error delimiter bits). Adding the 3-bit interframe gap, we have 23 bits. With a bus data rate of 1 Mbps (i.e., 1 μs bit time), the worst-case total error recovery time will be 23 μs. This is the worst-case time that the bus can recover, and the re-transmission of frames can start on the bus. This is rather a brief time and is one of the advantages of the CAN bus.

CAN bus error conditions will be described in detail in the next chapter.

3.4 Overload Frame

The overload frame is used by a receiving node to indicate that it is not yet ready to receive frames. As shown in Figure 3.9, the overload frame is similar to the error frame and consists of:

- 6-bits Overload Flag
- 8-bits Overload Delimiter

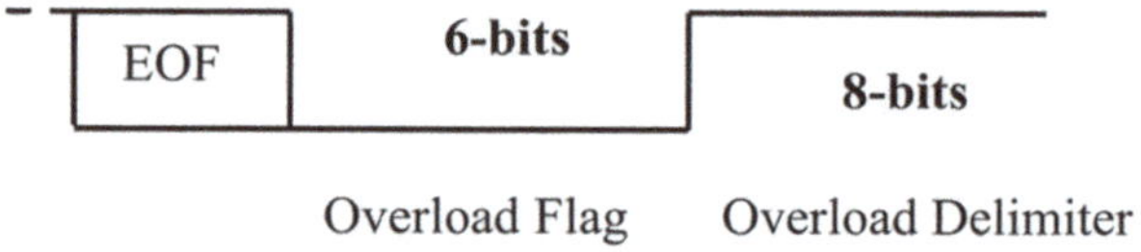

Figure 3.9: CAN Overload frame.

The overload frame can be sent between two data or remote frames. There are two kinds of overload conditions that can lead to the transmission of an overload frame:

1. The internal conditions of a receiver, which requires a delay of the next data frame or remote frame. i.e., a receiving node is temporarily overloaded and will not be able to receive any data, requiring a delay on the bus.

2. Detection of a dominant bit during the interframe gap or during the EOF frame.

The start of an overload frame due to case 1 is only allowed to be started at the first-bit time of an expected interframe gap. i.e., the 6-bit overload flag will override the 3 interframe gap bits. This is important and is the main timing difference between the error frame and the overload frame (remember that the error frame starts as soon as the error is detected where the frame is not completed and as such causes the frame to be discarded by all the nodes).

The overload frames due to case 2 start one bit after detecting the dominant bit. The Overload Flag consists of six dominant bits. The overload flag violates CAN bus rules (more than 5 bits of the same polarity on the bus) and as a consequence, all other nodes also detect an overload condition and on their part start transmission of an overload flag. As in the error flag, the overload flag can be extended to up to 12 bits. No more than two overload frames can be generated to delay a data or remote frame. The Overload Delimiter consists of eight recessive bits. The overload delimiter is of the same form as the error delimiter.

3.5 Extended CAN Frames

Now that we have explored all the standard CAN 2.0A frames, we can look at extended frames and see how the extended protocol differs from the standard protocol and also what additional benefits are available with the extended protocol.

CAN bus was originally developed for the automobile industry, and as such the standard message identifier is 11-bit. With 11 bits, it is possible to have 2048 different identifiers in a system. In order to improve the functionality of the bus and also increase the identifier capacity, the extended CAN (also called the CAN 2.0B) was introduced in 1985.

Figure 3.10 shows the data frame of Extended CAN.

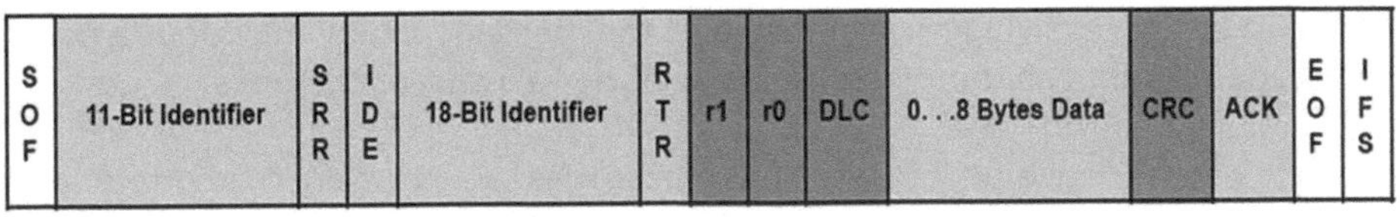

Figure 3.10: Extended CAN data frame.

As before, the SOF bit indicates the start of frame.

SRR replaces the RTR bit and is always recessive.

The IDE bit is transmitted as dominant for a CAN 2.0A bus, and as recessive for CAN 2.0B bus. The RTR bit is as in the standard protocol.

The other bits of the CAN 2.0B protocol are the same as the CAN 2.0A protocol.

In CAN 2.0B systems, the arbitration fields consist of all the fields from the 11-bit identifier to the RTR field. i.e., a total of 32 bits (see Figure 3.11). Since SRR and IDE are recessive bits in CAN 2.0B systems, in a mixed system where both the standard and the extended systems are used, a standard data frame will have a higher priority than an extended data frame.

The extended format comes with some tradeoffs:

- Longer bus latency times.
- Messages in extended format require about 20% more bandwidth.
- Decreased error detection performance due to more bits in a frame causing longer CRC calculation times.

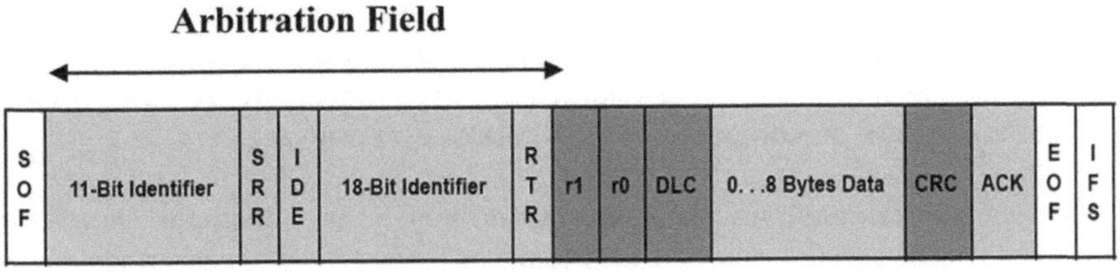

Figure 3.11: The arbitration field in extended CAN.

Chapter 4 • Data Exchange on the CAN Bus

4.1 Overview

In this chapter, you'll discover how data is exchanged on CAN bus. The CAN bus is a multi-master type where any node can be active transmitting or receiving at any one time.

When a node transmits on the bus, all other nodes listen and receive the transmitted message. Although all the receiving nodes receive all the messages, they may decide not to act on the message contents as the message may not be relevant to the node.

CAN bus nodes do not have addresses. Thus, messages are not transmitted to addressed nodes, but they are broadcast on the bus. Also, a node does not need to know where a message comes from. Similarly, a transmitting node may not necessarily know which nodes have actually acted on the message sent. The nodes only receive the messages they are programmed to receive.

Although the internal workings of different CAN controllers may be different, as far as data exchange mechanisms are concerned, the basic principles of data exchange are described briefly in this chapter.

4.2 Data exchange with data frames

Data frames are the most important frames in CAN bus, as they enable data to be sent from a transmitting node to other nodes on the bus. A data frame has a dominant RTR bit.

When a node wants to send data to other nodes on the bus, it forms a data frame. The data frame includes a message identifier, actual data bytes, and error checking bits. The message identifier is especially important as it is used by the receiving nodes to decide whether or not to accept this message. CAN bus controllers in receiving nodes have built in Acceptance Filters (or Receive Filters). These filters can be programmed and loaded with values by the programmer or system user. The message identifiers of data frames on the bus are compared with filter values, and the message is accepted by the controller if the filter value is the same as the message identifier. If the message identifier and the filter values are not the same, then the message is not accepted by the controller. Thus, by programming acceptance filters we can enable a node to accept or reject a message (note that all receiving nodes receive all messages, but they may not accept them all).

Most CAN bus controllers also have built in **Filter Masks**. These are used to determine which bits in the message identifier are to be compared with the values in the acceptance filters. For example, setting all filter mask values to 1s will make sure that all the bits of the message identifiers are to be compared with all bits of acceptance filters.

Figure 4.1 shows an example data exchange on the bus. In this example, there are three nodes on the bus, A, B, and C. Assume that the acceptance filters in each node's controller at the time of data transmission are set to the following values:

Node A: Acceptance filter value = "00000000011"
Node B: Acceptance filter value = "00000011111"
Node C: Acceptance filter value = "00000000111"

Assume that Node A has high priority and transmits a data frame with the message identifier set to bit pattern "00000000111". The other two nodes are in the receive mode and compare the message identifier on the bus with their acceptance filter values. Node C has the same acceptance filter value as the message identifier and thus its controller accepts the data frame sent by Node A. The controller in Node B compares its acceptance filter with the message identifier and ignores the data frame as the two are not the same.

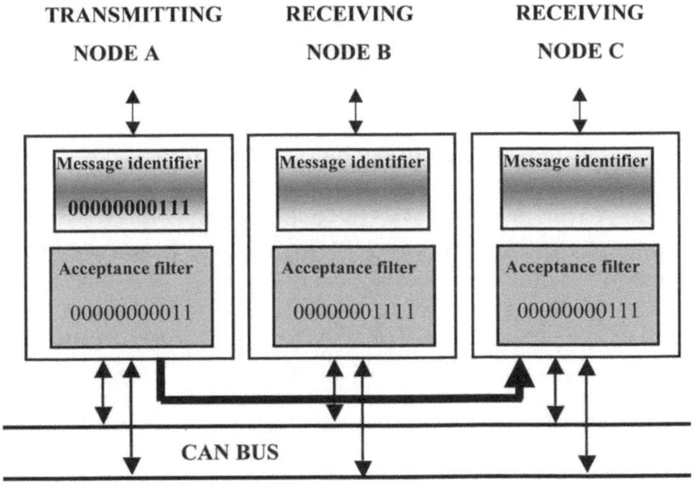

Figure 4.1: Data frame exchange on the bus.

Note that when a message is sent on the bus, the controllers of all nodes receive this message, but may not accept the message. Also, although the controller may accept a message, this message usually stays in a receive buffer of the controller until the application software makes a request to pull the message out of this buffer and copy it to its internal data structures.

Figure 4.2 shows another example where Node A transmits a data frame and both Node B and Node C accept this data frame. In this example, the acceptance filter of Node B is changed to "00000000111".

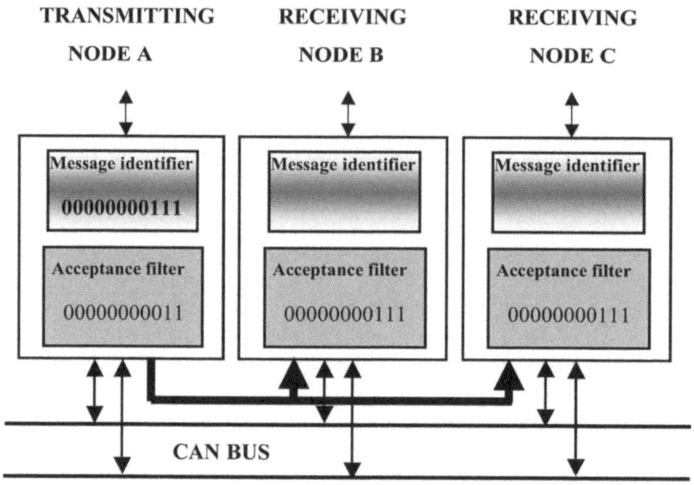

Figure 4.2: Example data frame exchange.

In Figure 4.3, Node A transmits a data frame but neither of the other two nodes accept this data frame. Here, the acceptance filters of Node B and Node C are changed to "00000111111" and "11110000000" respectively.

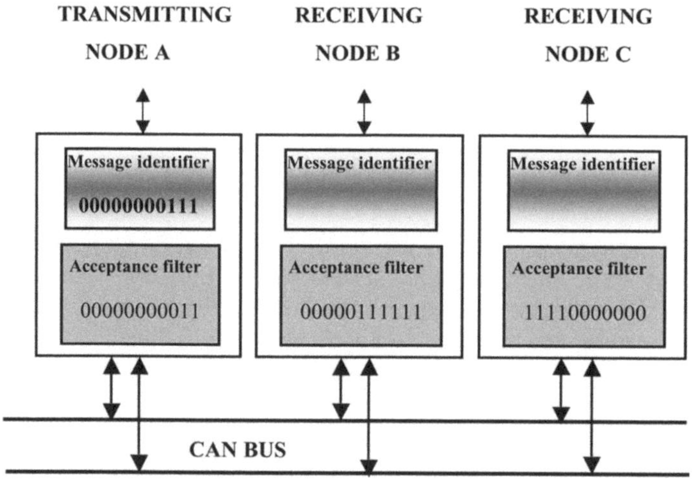

Figure 4.3: Example data frame exchange.

The data exchange mechanism when a data frame is sent on the bus is summarized below:

- A node transmits a data frame with a certain message identifier.
- The controllers of all nodes on the bus receive this message.
- The controllers of the receiving nodes on the bus compare this message identifier with their acceptance filters.

- The nodes whose acceptance filters match the message identifier accept the data frame. The acceptance filters in these nodes were set such that the message with the specified message identifier is relevant to the application layers of these nodes.

- The nodes whose acceptance filters do not match the message identifier ignore the data frame. The acceptance filters in these nodes were set such that the message with the specified message identifier is not relevant to the application layers of these nodes.

Note that message identifiers and acceptance filter values are set by the engineers designing and maintaining the bus.

4.3 Remote frames on the bus

Remote frames are recognized by a recessive RTR (Remote Transmission request) bit in their frames and not having data fields. These frames are used to request data from another node. When a remote frame is sent on the bus, another node sends the requested data in a data frame. Both of these nodes have the same message identifiers.

Figure 4.4 shows an example where three nodes are on the bus with the message identifiers:

Node A: Acceptance filter value = "00000000011"
Node B: Acceptance filter value = "00000011111"
Node C: Acceptance filter value = "00000000111"

Node A sends a remote request by setting the RTR bit to recessive state and the identifier to "00000000111". Nodes B and C receive the remote request, and Node C accepts this remote request.

In Figure 4.5, Node C sends the requested data with a data frame. In this example, Node A receives the requested data frame. Note that Node A sends the requested data frame with the same identifier as the remote frame.

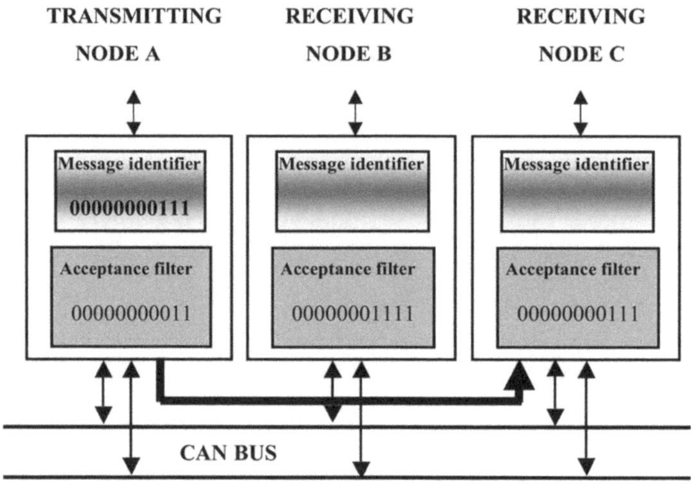

Figure 4.4: Example of remote frame.

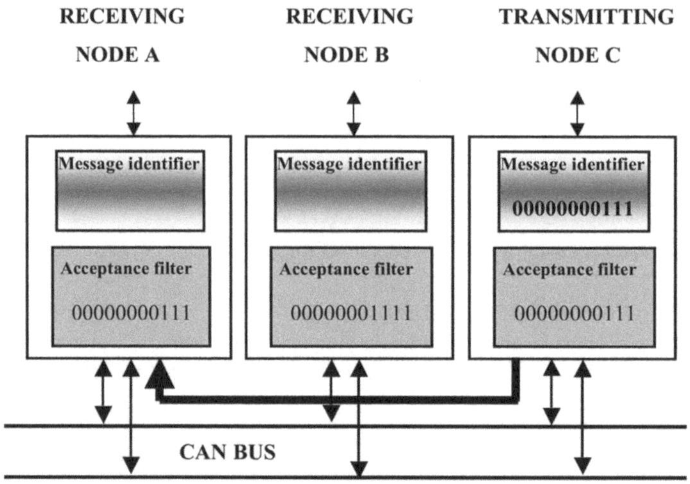

Figure 4.5: Example data frame with requested data.

Chapter 5 • CAN Bus Interface Modules

5.1 Overview

In this chapter, join us in exploring the CAN bus interface module that will be used in the projects in this book.

5.2 MCP2515 CAN bus interface module

There are many CAN bus modules available in the market from various manufacturers. In this book, we will be using the **Youmile MCP2515** pictured in Figure 5.1.

Figure 5.1: Youmile MCP2515 CAN bus interface module.

The module measures 1.57 inches by 1.1 inches (4 mm × 2.8 mm), with a 2-pin screw terminal block at one end of the board to connect to the CAN bus. The other end of the board has a 7-pin header with the following pin definitions:

INT: interrupt
SCK: clock
SI: SPI MOSI
SO: SPI MISO
CS: chip select
GND: ground
VCC: Power (+5 V)

Figure 5.2 shows the operational block diagram of the CAN bus interface module.

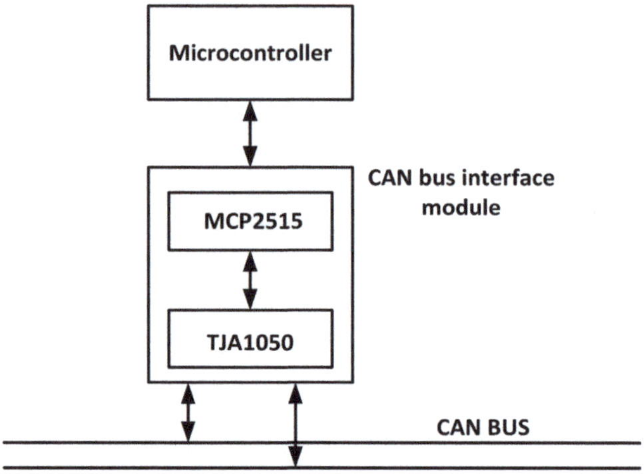

Figure 5.2: Operational block diagram.

Figure 5.3 shows the circuit diagram of the interface module. The CAN bus interface module contains two chips: CAN Controller (MCP2515), and CAN Transceiver (TJA1050).

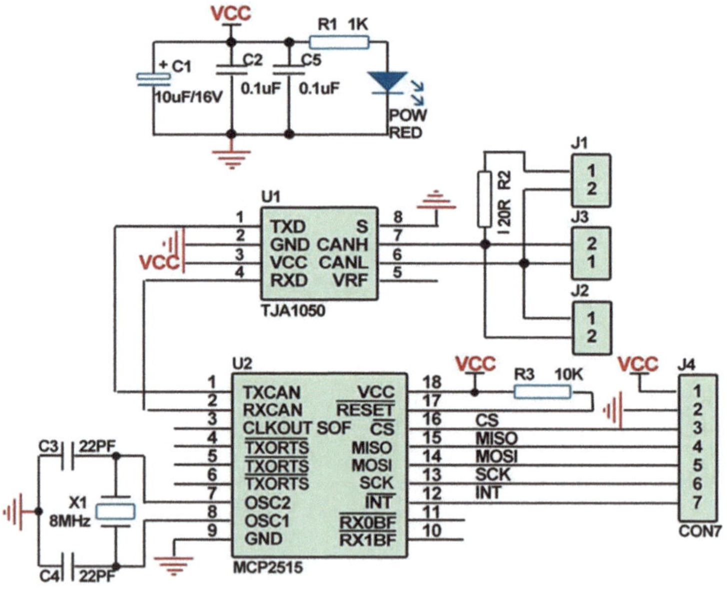

Figure 5.3: Circuit diagram of the interface module.

The board operates with the standard SPI bus and has the following features:

- MCP2515 CAN controller chip
- TJA1050 CAN Transceiver chip
- Operation with +5 V power supply
- 5 mA working current (1 microamp standby current)

- On-board 120-ohm terminating resistors
- Operation with on-board 8-MHz crystal
- Support of CAN V2.0B protocol
- 1 Mb/s communication speed
- 0-8 byte data field
- Working temperature: –40 ºC to 85 ºC industrial grade

It is worthwhile to look at the details of CAN bus interface module used in the projects in this chapter.

5.2.1 The MCP2515 CAN controller chip

This is a 1-Mbps controller compatible with CAN 2.0B protocol specification. This chip has the following specifications:

- Two receive buffers
- Three transmit buffers
- Two acceptance masks
- Six acceptance filters
- SPI interface to the microcontroller
- Output clock pin
- Interrupt pin
- SOF signal is available for monitoring
- Low-power operation

The chip contains the following modules:

- CAN controller
- Control logic
- SPI interface logic
- Control and interrupt registers

The MCP2515 chip can operate from +3.3 V to +5 V. The CAN controller module consists of the CAN protocol engine, TX and RX buffers, acceptance masks and filters. Messages detected on the bus are checked for errors. If there are no errors, the message identifier is compared with the user-defined filters to see if there is a match. If a match is detected, the message is accepted by the module and is ready to be read by the microcontroller. The CRC and the bit timing are also controlled by the CAN controller module.

Communication between the microcontroller and the MCP2515 employs the SPI protocol, and this is handled by the SPI interface logic.

The control logic is responsible for controlling all internal operations of the chip.

Control and interrupt registers are under the control of the control logic, and they provide interrupt capability to the chip.

The MCP2515 chip is controlled from the Serial Peripheral Interface (SPI) bus available on many microcontrollers. Commands and data are sent to the chip via the SI pin, with the rising edge of the clock CLK input. Similarly, data is out from pin SO on the falling edge of the clock. The chip select input, CS, must be low for normal operation of the device. The registers in the device must be programmed before the device can be used in a microcontroller circuit.

As shown in Figure 5.4, the MCP2515 includes two receive buffers with multiple acceptance filters. There is also a separate Message Assembly Buffer (MAB) that acts as a third buffer, where this buffer receives the text message from the bus. MAB assembles all messages received and these messages are transferred to the RXB0 and RXB1 buffers only if the acceptance filter criteria is met. Messages received in the MAB are initially applied to the mask and filters of RXB0. In addition, only one filter match occurs (e.g., if the message matches both RXF0 and RXF2, the match will be for RXF0 and the message will be moved into RXB0). The MCU accesses one buffer, while the other buffer is available for message reception or for holding a previously received message. RXB0 is the higher-priority buffer and has one mask and two acceptance filters. RXB1 is the lower-priority buffer, with one mask and four acceptance filters associated with it. Message is applied to the RB0 mask and filters first. RX0BF and RX1BF are Receive-Buffer-Full pins, and they can be used to indicate that a valid message has been loaded into RXB0 or RXB1 respectively. INT pin provides an interrupt to the MCU for many different conditions. When an interrupt occurs, we should check the nature of this interrupt whether it is because data is received, an error condition occurred, or some other condition.

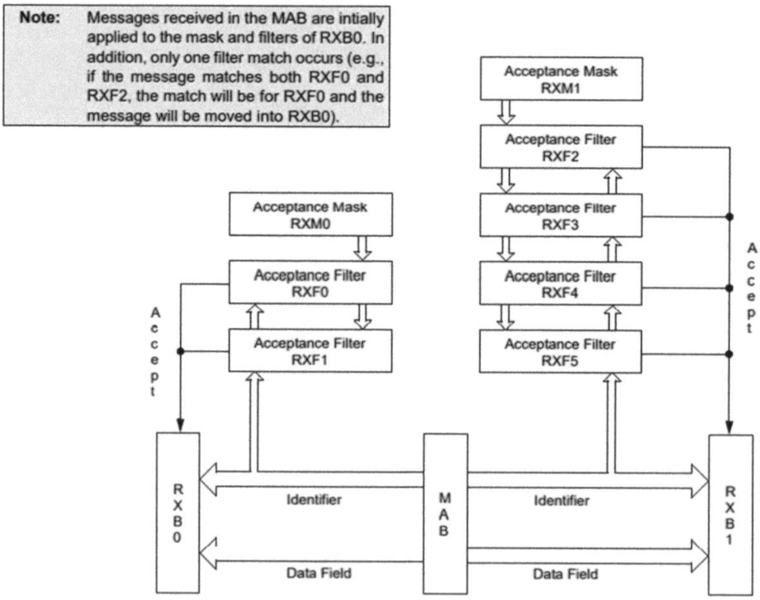

Figure 5.4: MCP2515 receive buffers.

The message acceptance filters and masks are used to determine if a message in the Message Assembly Buffer (MAB) should be loaded into either of the receive buffers. Once a valid message has been received into the MAB, the identifier fields of the message are compared to the filter values. If there is a match, that message will be loaded into the appropriate receive buffer. The truth table given in Table 5.1 shows the mask and filter matching algorithm.

Mask bit n	Filter bit n	Message identifier bit	Accept or reject
0	x	x	Accept
1	0	0	Accept
1	0	1	Reject
1	1	0	Reject
1	1	1	Accept

Table 5.1: Mask and filter match truth table.

When a mask bit is 0, the filter and message IDs are not checked. Examples of setting the acceptance masks and filters are given below.

Example
It is required to receive CAN frames from nodes 1, 2, 3 and 4. What are the required mask and filter values?

Solution
Write down binary values of the required IDs:

ID1 = 00000000001
ID2 = 00000000010
ID3 = 00000000011
ID5 = 00000000100

Set the filter mask to ignore the bits of the ID that are common with all the wanted IDs. i.e., set the bits of the mask register to 1 if the ID bits are the same. We therefore have:

MASK: 11111111000

The above mask will cause the filter to ignore the three LSB bits which are not common. Now, set the filter to match any of the common IDs. e.g., 00000000001

The required settings are, therefore:

MASK: 11111111000
FILTER: 00000000001

Example

It is required to receive CAN frames from ID 1 only. What will be the filter and mask values?

Solution

ID: 1 = 00000000001

The required filter setting is: 00000000001
The required mask setting is: 11111111111

With all 1s, each bit of the filter will be used to check incoming IDs.

5.2.2 The TJA1050 CAN transceiver chip

The CAN Transceiver chip is the interface chip between the CAN controller chip and the physical CAN bus wires. The basic features of this chip are:

- +5 V operation
- compatible with the ISO 11898 standard
- differential receiver
- unpowered node does not disturb the bus lines
- transmit data dominant time-out function
- bus pins are protected against transients in an automotive environment
- input levels compatible with both 3.3-V and 5-V devices
- thermally protected
- at least 110 nodes can be connected
- 50 mA supply current in dominant mode, 5 mA in recessive mode

In the next chapters, you'll uncover the use the CAN bus interface module in Arduino Uno and Raspberry Pi-based projects.

Chapter 6 • Arduino Uno CAN Bus Projects

6.1 Overview

This chapter presents the step-by-step development of projects using Arduino Uno development boards connected over a CAN bus through CAN bus interface modules.

6.2 Arduino Uno CAN bus interface

The CAN bus interface module is controlled by the microcontroller SPI bus. The Arduino Uno has the following SPI bus pins (Figure 6.1):

MOSI	pin 11
MISO	pin 12
SCK	pin 13
SS	pin 10

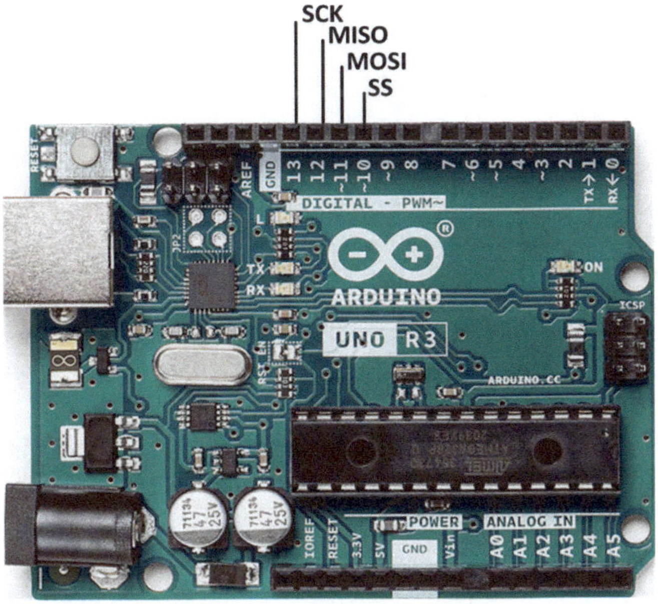

Figure 6.1: Arduino Uno SPI pins.

6.3 Project 1: Simple Arduino to Arduino CAN bus communication

Description: In this simple project, you name two Arduino Unos as the **GENERATOR** and the **SQUARER** and connect them on a CAN bus (i.e., you have two nodes called **GENERATOR** and **SQUARER**). The **GENERATOR** node will generate random integer numbers between 1 and 20 and send them to the **SQUARER** node, where the received numbers will be squared and displayed on the Serial Monitor of the IDE. The aim of this project is to show how two Arduino Unos can be connected on a CAN bus.

Block diagram: Figure 6.2 shows the block diagram of the project showing two Arduino Unos, two CAN bus interface modules, and the CAN bus wire.

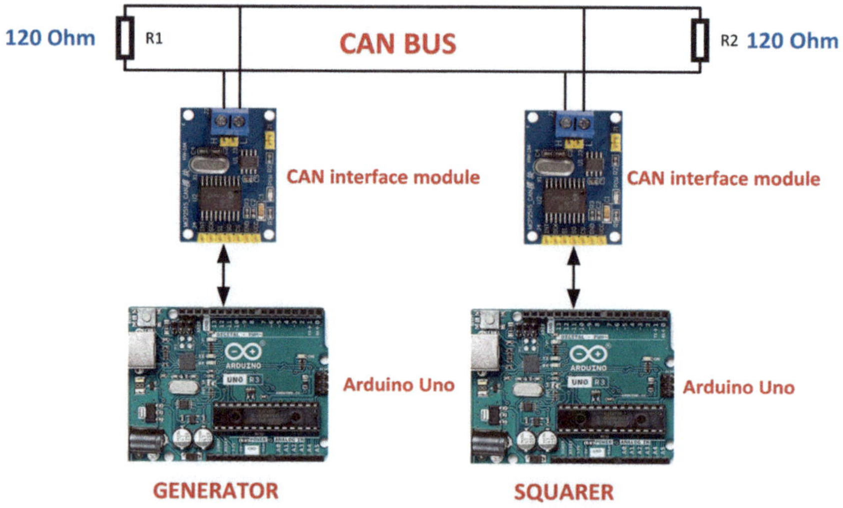

Figure 6.2: Block diagram of the project.

Circuit diagram: Figure 6.3 shows the circuit diagram of the project. Notice that connection to the CAN bus is made through screw terminal J2 so that the on-board 120-ohm resistor is not used (you could have used the onboard terminating resistors by shorting jumper J1 and connecting the bus wires to jumper J3 instead of J2). Because both the Arduino Uno and CAN interface module are +5 V compatible, there is no need to use any voltage-level converter chips. In this project, about a 1-meter twisted pair cable was used with a 120-ohm terminator resistor at each end.

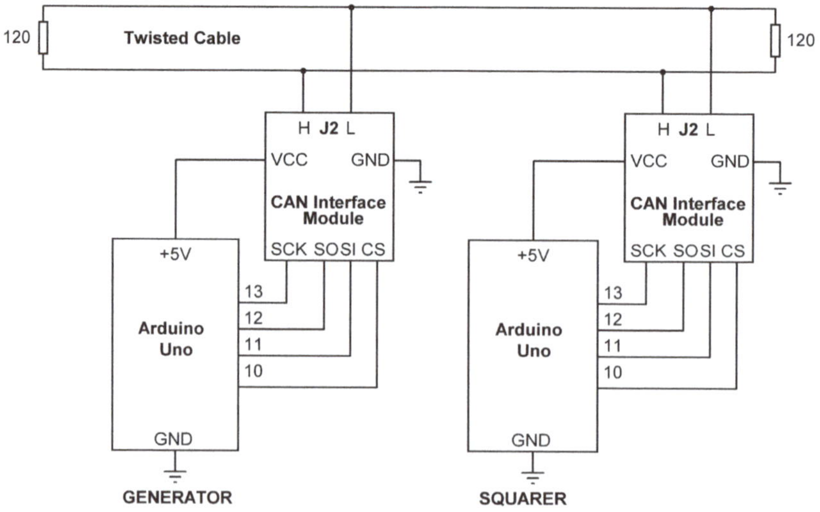

Figure 6.3: Circuit diagram of the project.

CAN bus library

Before writing the programs for the **Generator** and **Squarer**, you have to install a CAN bus library to your Arduino IDE. There are several CAN bus libraries available on the Internet. The one used in this book, in the Arduino-based projects is the MCP2515 CAN bus library, available at the following website (a copy is available in the software support archive for the book):

https://github.com/autowp/arduino-mcp2515/

The steps to install the library in the Arduino IDE are:

- Download the ZIP file from the following site:

  https://github.com/autowp/arduino-mcp2515/archive/master.zip

- Start your Arduino IDE.
- Click **Sketch → Include Library → Add ZIP Library...** and browse to the library zip file you just downloaded.
- Restart your Arduino IDE.
- You should see example files in the Examples folder.

You should include the following two header files at the beginning of your programs:

```
#include <SPI.h>
#include <mcp2515.h>
```

CAN bus library functions

By default, it is assumed that the SPI CS is connected to pin 10 of the Arduino Uno. For example, the following function is used to specify the SPI CS pin:

```
MCP2515 mcp2515(10);
```

Operational modes

The following operational modes are available:

```
mcp2515.setNormalMode();
mcp2515.setLoopbackMode();
mcp2515.setListenOnlyMode();
```

The following function is used to set to Normal mode:

```
mcp2515.setNormalMode();
```

Loop-back is a feature that allows the CAN controller to communicate with itself without actually sending anything out on the bus.

In Listen Only Mode, the CAN module does not acknowledge any messages on the bus, but it will read the messages silently. This mode is useful in bus monitoring applications.

Baud rates

The following baud rates are available:

```
enum CAN_SPEED {
    CAN_5KBPS,
    CAN_10KBPS,
    CAN_20KBPS,
    CAN_31K25BPS,
    CAN_33KBPS,
    CAN_40KBPS,
    CAN_50KBPS,
    CAN_80KBPS,
    CAN_83K3BPS,
    CAN_95KBPS,
    CAN_100KBPS,
    CAN_125KBPS,
    CAN_200KBPS,
    CAN_250KBPS,
    CAN_500KBPS,
    CAN_1000KBPS
};
```

The following function is used to set the baud rate to 125 kB/s:

```
mcp2515.setBitrate(CAN_125KBPS);
```

Clock speed

The default clock speed is 16 MHz. Available clock speeds are:

```
enum CAN_CLOCK {
    MCP_20MHZ,
    MCP_16MHZ,
    MCP_8MHZ
}
```

CANframe

The CAN frame contains the following elements:

```
struct can_frame {
    uint32_t can_id;  /* 32 bit CAN_ID + EFF/RTR/ERR flags */
    uint8_t  can_dlc;
    uint8_t  data[8];
};
```

Sending data

The following function is used to send data on the bus:

```
mcp2515.sendMessage(&frame);
```

where for example, the frame can be the following structure with standard frame (can_id = 0x0000) and 4 bytes of data to be sent:

```
struct can_frame frame;
frame.can_id = 0x0000;
frame.can_dlc = 4;
frame.data[0] = 0xFF;
frame.data[1] = 0xFF;
frame.data[2] = 0xFF;
frame.data[3] = 0xFF;
```

Receiving data

The following function is used to receive data on the bus:

```
mcp2515.readMessage(&frame)
```

Where the frame is in the send function. Notice that this function can only get frames that meet the requirements of masks and filters.

Receive Masks and Receive Filters

There are 2 receive mask registers and 5 receive filter registers (also called **Acceptance Filters**) on the controller chip that guarantee you get data from the target device. They are useful especially in a large network consisting of numerous nodes.

MASK mask represents one of two masks. For example, **MCP2515::MASK0** represents MASK1 and **MCP2515::MASK2** represents MASK2.

RXF num represents one of six acceptance filters registers from **MCP2515::RXF0** to **MCP2515::RXF5**.

ext represents the status of the frame. false means it's a mask or filter for a standard frame. true means it's for an extended frame.

ulData represents the content of the mask of filter.

Node: GENERATOR program listing:

Figure 6.4 shows the program listing (Program: **GENERATOR**). The required header files are included at the beginning of the program. Then, the **MyMSg** structure is defined to store the CAN frame data, and SPI CS is set to pin 10 of the Arduino.

Inside the **setup()** function, the SPI bus is started, MCP2515 is reset, the clock frequency of the interface is set to 8 MHz with a baud rate of 500 kB/s. Also, the MCP2515 is set to operate in normal mode.

Inside the main program loop, the CAN bus ID (**can_id**) is set to 0x30 and the message byte count (**can_dlc**) is set to 8 (although the first byte is used). A random integer number is generated between 1 and 20 (inclusive) and this number is sent over the CAN bus using the function **sendMessage()**. This process is repeated after 5 seconds of delay, where a new random number is generated and sent to node **SQUARER**.

```
/*----------------------------------------------------------
                    GENERATOR
                    =========

This program generates random numbers between 1 and 20 and
sends them to the SQUARER program over the CAN bus. The
SQUARER program takes the square of the received numbers
and then displays them on the Serial Monitor

Author: Dogan Ibrahim
File  : GENERATOR
Date  : December 2022
----------------------------------------------------------*/
#include <SPI.h>
#include <mcp2515.h>

#define CS 10
int RandomNumber;

struct can_frame MyMsg;
MCP2515 mcp2515(CS);                         // SPI CS

void setup()
{
  SPI.begin();
  mcp2515.reset();
  mcp2515.setBitrate(CAN_500KBPS, MCP_8MHZ);  // CAN bus config
  mcp2515.setNormalMode();                    // Normal mode
}

void loop()
{
  RandomNumber = random(1, 21);               // Random number
  MyMsg.can_id = 0x30;                        // CAN ID
  MyMsg.can_dlc = 8;                          // 8 bytes
  MyMsg.data[0] = RandomNumber;               // Number in data[0]
```

```
    MyMsg.data[1] = 0x00;                         // Remainders 0
    MyMsg.data[2] = 0x00;
    MyMsg.data[3] = 0x00;
    MyMsg.data[4] = 0x00;
    MyMsg.data[5] = 0x00;
    MyMsg.data[6] = 0x00;
    MyMsg.data[7] = 0x00;

    mcp2515.sendMessage(&MyMsg);                  // Send message
    delay(5000);                                  // Wait 5 secs
}
```

Figure 6.4: Program: **GENERATOR**.

Node: SQUARER program listing

Figure 6.5 shows the program listing (Program: **SQUARER**). The beginning and **setup()** function of this program is similar to Figure 6.4, but here the Serial Monitor is initialized at 9600 baud.

Inside the main program loop, the program checks if there are any messages on the CAN bus and reads these messages. In this program, a message is an integer number between 1 and 20. The received number and its square are both displayed on the Serial Monitor. Notice that receive masks and receive filters (also called **Acceptance Filters**) are not used in this program. Without receive masks and filters, all the nodes on the bus will receive the transmitted messages.

```
/*-----------------------------------------------------------
                    SQUARER
                    =======

This program receives random numbers between 1 and 20 and
takes the square of these numbers and then displays the
numbers on the Serial Monitor

Author: Dogan Ibrahim
File  : Squarer
Date  : December 2022
-----------------------------------------------------------*/
#include <SPI.h>
#include <mcp2515.h>

#define CS 10
int ReceivedNumber;

struct can_frame MyMsg;
MCP2515 mcp2515(CS);                              // SPI CS
```

```
void setup()
{
  Serial.begin(9600);                              // Serial Monitor
  Serial.println("CAN BUS");
  SPI.begin();
  mcp2515.reset();
  mcp2515.setBitrate(CAN_500KBPS, MCP_8MHZ);       // CAN config
  mcp2515.setNormalMode();                         // Normal mode
}

void loop()
{
  if(mcp2515.readMessage(&MyMsg) == MCP2515::ERROR_OK)
  {
    ReceivedNumber = MyMsg.data[0];
    Serial.print("Received Number is: ");
    Serial.println(ReceivedNumber);
    Serial.print("Square is: ");
    Serial.println(ReceivedNumber * ReceivedNumber);
    Serial.println("");
  }
}
```

Figure 6.5: Program: **SQUARER**.

Testing the project: Construct the project as shown in the circuit diagram and power up both Arduino Unos. Start the Serial Monitor on the **SQUARER** node and you should see the received numbers and their squares displayed every 5 seconds. A sample output is shown in Figure 6.6. The text **CAN BUS** is displayed at the beginning of the output.

```
COM8

CAN BUS
Received Number is: 8
Square is: 64

Received Number is: 10
Square is: 100

Received Number is: 14
Square is: 196

Received Number is: 19
Square is: 361

Received Number is: 11
Square is: 121
```

Figure 6.6: Sample output on Serial Monitor.

Constructing the project: Figure 6.7 shows the project with two Arduino Unos, two CAN bus interface modules, and the CAN bus. Notice that the INT pin of the CAN bus interface modules is not used.

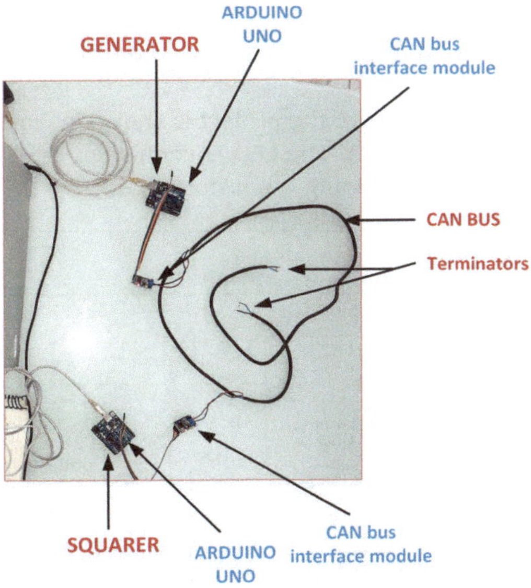

Figure 6.7: Constructing the project.

6.4 Project 2: Pushbutton and LED

Description: In this project, two Arduino Unos communicate over the CAN bus, similar to the previous project. Here, a pushbutton is connected to node **SENDER** and an LED is connected to node **RECEIVER**. Pressing the pushbutton turns on the LED for 5 seconds at the **RECEIVER** node.

Block diagram: Figure 6.8 shows the block diagram of the project.

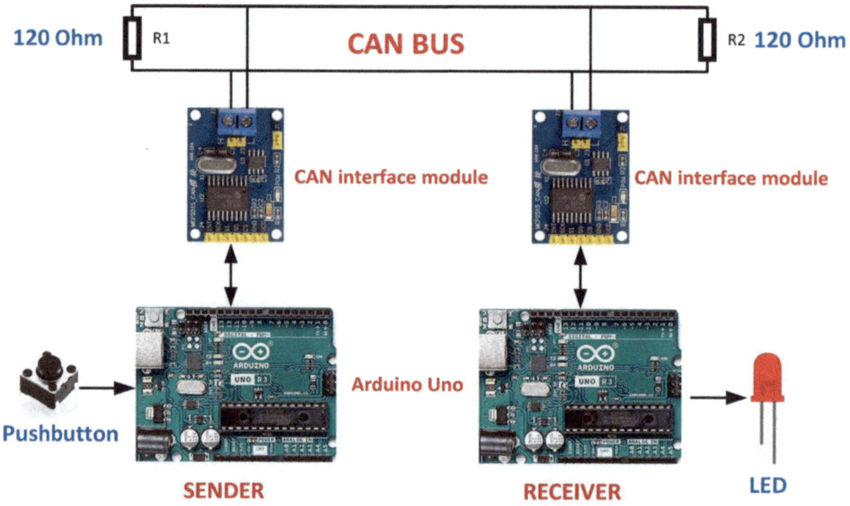

Figure 6.8: Block diagram of the project.

Circuit diagram: The circuit diagram of the project is shown in Figure 6.9. The pushbutton is connected to pin 2 of node **SENDER**. The LED is connected to pin 2 of node **RECEIVER** through an external 1-kohm current-limiting resistor.

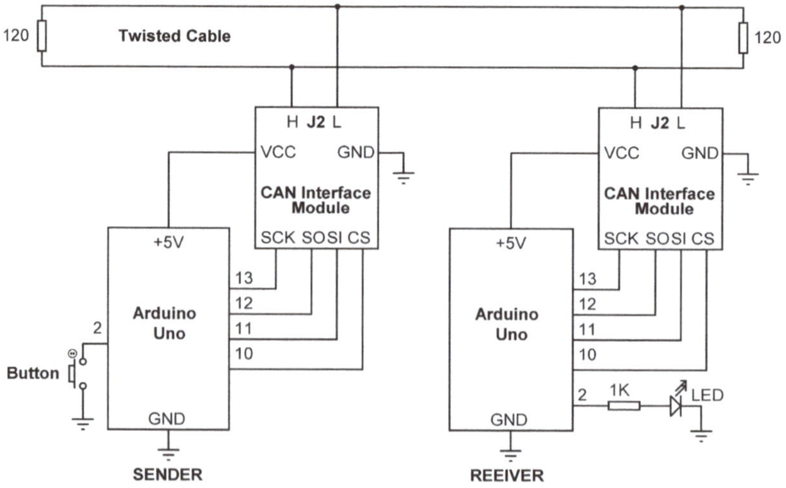

Figure 6.9: Circuit diagram of the project.

Node: SENDER program listing

Figure 6.10 shows the program listing (Program: **SENDER**). The beginning parts of the program are similar to the program shown in Figure 6.4. Here, pin 2, where the pushbutton is connected, is configured as an input port with the internal pull-up resistor enabled so that the default state of the button is at logic 1. Inside the main program loop, the state of the pushbutton is checked. If the pushbutton is pressed (i.e., **Button** is 0), then node **SENDER** transmits number 1 over the CAN bus, otherwise, 0 is transmitted. When node **RECEIVER** receives 1 it turns ON the LED for 5 seconds. This process is repeated forever until stopped by the user.

```
/*----------------------------------------------------------
                    SENDER
                    ======

In this program a button is connecte to pin 2 of node
SENDER. Pressing the button sends logic 1 to node RECEIVER
where the LED at node RECEIVER is turned on for 5 seconds.
The button is pulled-up in software

Author: Dogan Ibrahim
File  : SENDER
Date  : December 2022
----------------------------------------------------------*/
#include <SPI.h>
#include <mcp2515.h>

#define CS 10
int Button= 2;

struct can_frame MyMsg;
MCP2515 mcp2515(CS);

void setup()
{
  SPI.begin();
  mcp2515.reset();
  mcp2515.setBitrate(CAN_500KBPS, MCP_8MHZ);
  mcp2515.setNormalMode();
  pinMode(Button, INPUT_PULLUP);           // Pullup button
}

void loop()
{
  while(digitalRead(Button) == 1);         // Wait if not pressed
  while(digitalRead(Button) == 0);         // Wait until released
  MyMsg.can_id = 0x30;
```

```
    MyMsg.can_dlc = 8;
    MyMsg.data[0] = 1;                        // Load 1
    MyMsg.data[1] = 0x00;
    MyMsg.data[2] = 0x00;
    MyMsg.data[3] = 0x00;
    MyMsg.data[4] = 0x00;
    MyMsg.data[5] = 0x00;
    MyMsg.data[6] = 0x00;
    MyMsg.data[7] = 0x00;

    mcp2515.sendMessage(&MyMsg);              // Send message
    delay(10);
}
```

Figure 6.10: Program: **SENDER**.

Node: RECEIVER program listing

Figure 6.11 shows the program listing (Program: **RECEIVER**). Port 2, where the LED is connected, is configured as an output and the LED is turned OFF at the beginning of the program. When number 1 is received over the CAN bus, the LED is turned ON for 5 seconds, and then turned OFF.

```
/*----------------------------------------------------------
                    RECEIVER
                    ========

In this program an LED is connected to port pin 2. The LED
is turned ON for 5 seconds when the button at node SENDER
is pressed

Author: Dogan Ibrahim
File  : RECEIVER
Date  : December 2022
----------------------------------------------------------*/
#include <SPI.h>
#include <mcp2515.h>

#define CS 10
int ReceivedMessage;
int LED = 2;                                 // LED at port 2

struct can_frame MyMsg;
MCP2515 mcp2515(CS);

void setup()
{
```

```
  SPI.begin();
  mcp2515.reset();
  mcp2515.setBitrate(CAN_500KBPS, MCP_8MHZ);  // Config CAN bus
  mcp2515.setNormalMode();                     // Normal mode
  pinMode(LED, OUTPUT);                        // LED is output
  digitalWrite(LED, 0);                        // LED OFF
}

void loop()
{
  if(mcp2515.readMessage(&MyMsg) == MCP2515::ERROR_OK)
  {
    ReceivedMessage = MyMsg.data[0];           // Received data
    if(ReceivedMessage == 1)                   // If 1 is received
    {
      digitalWrite(LED, 1);                    // Turn ON LED
      delay(5000);                             // Wait 5 seconds
      digitalWrite(LED, 0);                    // Turn OFF LED
    }
    else
      digitalWrite(LED, 0);
  }
}
```

Figure 6.11: Program: **RECEIVER**.

Testing the project: Construct the project as shown in the circuit diagram and power up both Arduino Unos. Press the pushbutton, and you should see the LED to be turned ON for 5 seconds.

Constructing the project: Figure 6.12 shows the project constructed on two breadboards, one for each node.

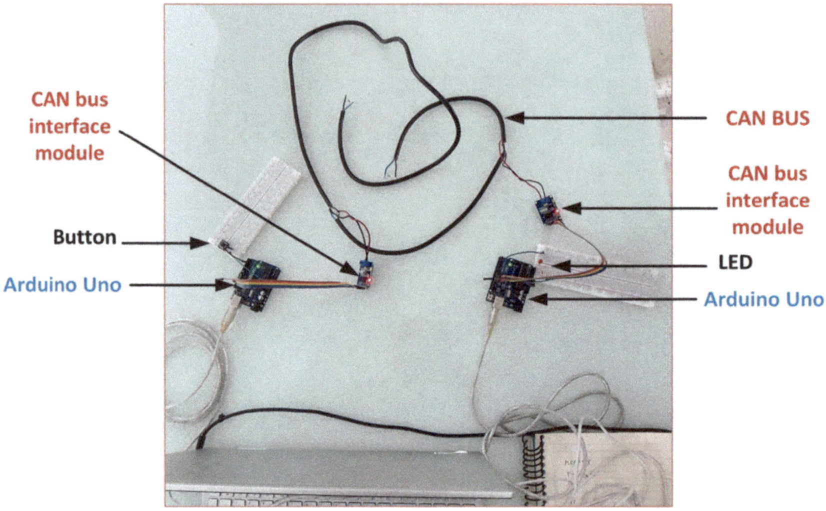

Figure 6.12: Constructing the project.

6.5 Project 3: Pushbutton and LED with CAN bus interrupts

Description: This project is similar to the previous project, but here CAN bus interrupts are used to receive data instead of polling the bus.

Block diagram: The block diagram of the project is the same as in Figure 6.8, but node **RECEIVER** is named as **RECEIVERINT**.

Circuit diagram: The circuit diagram is shown in Figure 6.13. It is very similar to Figure 6.9, but here pin 3 of node **RECEIVER** (called **RECEIVERINT** in this project) is used as an external interrupt pin and is connected to the **INT** pin of the CAN bus interface module. It is important to remember that pin 2 and pin 3 of Arduino Uno can be used as external interrupt pins.

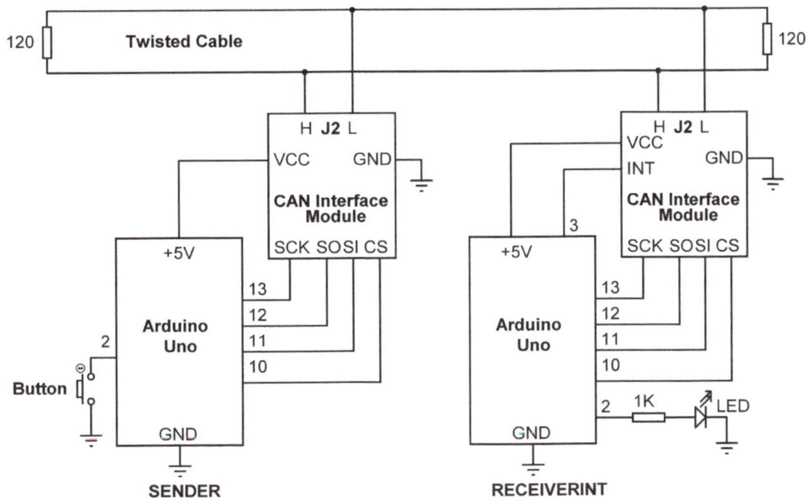

Figure 6.13: Circuit diagram of the project.

Sender Arduino Uno program listing

This program is exactly the same as the one in Figure 6.10 (i.e., Program: **Sender**).

ReceiverInt Arduino Uno program listing

Figure 6.14 shows the program listing (Program: **ReceiverInt**). Port 2, where the LED is connected, is configured as an output and the LED is turned OFF by default. Port 3 is used as an external interrupt pin, and function **attachInterrupt()** is used to attach pin 3 to CAN bus interrupts with the interrupt service routine named function **MyISR()**. When data is received by the CAN bus interface module, an interrupt is generated at pin 3 of Arduino Uno. The program then jumps to function **MyISR()**. Inside this function, the type of interrupt is checked, and if it is from receive buffer **RX0IF** then the message is read and variable **flag** is assigned to the first byte of the message. Therefore, **flag** is either 0 or 1. If **flag** is 1, then the LED is turned ON for 5 seconds inside the main program loop.

Note that when a message is moved into either of the receive buffers **RXB0** or **RXB1**, the appropriate **RXnIF** bit (**CANINTF**) is set. It is advisable not to have a delay code inside an interrupt service routine.

```
/*----------------------------------------------------------
                     RECEIVERINT
                     ===========

This program receives command from node SENDER. The program
is interrupt driven. When a message is received over the CAN
bus, an external interrupt is generated where the command is
read.If the comamnd is 1 then the LED is turned ON for 5 seconds

Author: Dogan Ibrahim
```

```
File   : RECEIVERINT
Date   : December 2022
----------------------------------------------------------*/
#include <SPI.h>
#include <mcp2515.h>

#define CS 10
int LED = 2;
int InterruptPin = 3;
int irq;
volatile int flag = 0;

struct can_frame MyMsg;
MCP2515 mcp2515(CS);

//
// Interrupt service routine
//
void MyISR()
{
  irq = mcp2515.getInterrupts();                    // Get type of interrupt
  if(irq & MCP2515::CANINTF_RX0IF)                  // RX0IF ?
  {
    if(mcp2515.readMessage(MCP2515::RXB0, &MyMsg) == MCP2515::ERROR_OK)
    {
      mcp2515.clearInterrupts();                    // Clear MCP2515 interrupts
      flag = MyMsg.data[0];                         // Assign to flag
    }
  }
}

//
// CAN interfce module configuration and interrupt config on pin 3
//
void setup()
{
  SPI.begin();
  mcp2515.reset();
  mcp2515.setBitrate(CAN_500KBPS, MCP_8MHZ);        // CAN bus configure
  mcp2515.setNormalMode();                          // Normal mode
  pinMode(LED, OUTPUT);                             // LED is output
  digitalWrite(LED, 0);                             // LED is OFF
  pinMode(InterruptPin, INPUT_PULLUP);
  attachInterrupt(digitalPinToInterrupt(InterruptPin), MyISR, FALLING);
}
```

```
void loop()
{
  if(flag == 1)                          // If interrupt and 1 received
  {
    flag = 0;
    digitalWrite(LED, 1);                // LED ON
    delay(5000);                         // 5 seconds delay
    digitalWrite(LED, 0);                // LED OFF
  }
}
```

Figure 6.14: Program: **RECEIVERINT**.

Testing the project: Construct the project as shown in the circuit diagram and power up both Arduino Unos. Press the pushbutton, and you should see the LED be turned ON for 5 seconds.

6.6 Project 4: Remote temperature alarm

Description: This project consists of two nodes, called **SENSOR** and **ALARM**. At node **SENSOR** there is a temperature sensor, which measures the ambient temperature and sends messages to node **ALARM**. If the temperature rises above a preset value (e.g., 30 ºC) then a message is sent to node **ALARM** to indicate this fact. At node **ALARM** there is an active buzzer. The buzzer is activated when the received message indicates that the temperature has risen above the preset value. This process is repeated continuously.

Block Diagram: The block diagram of the project is shown in Figure 6.15.

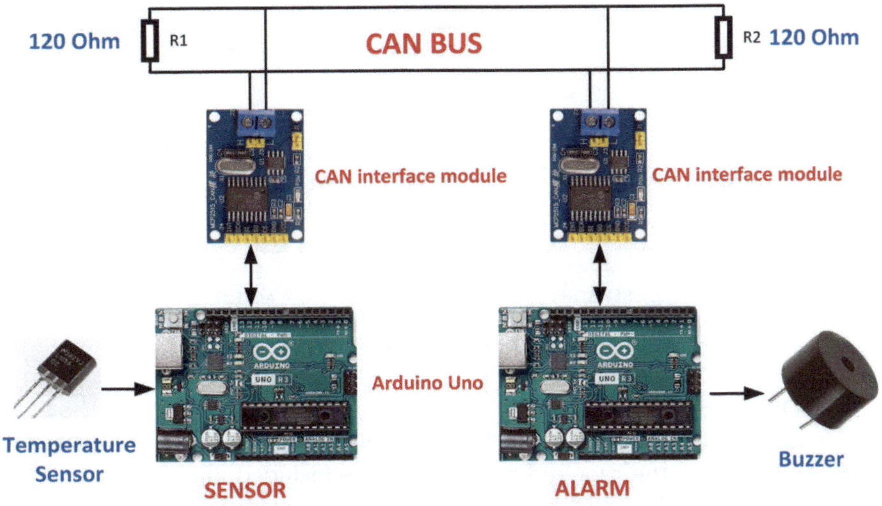

Figure 6.15: Block diagram of the project.

Circuit Diagram: In this project, the ambient temperature is measured using an LM-35DZ-type analog temperature sensor chip (Figure 6.16), connected to analog input A0 of node **SENSOR**. LM35DZ sensor chip has 3 pins: power supply (Vs), ground (GND), and output (Vout). The output voltage is proportional to the measured voltage and is given by Vo = 10 mV/ºC. Thus, for example, at 20 ºC the output voltage is 200 mV, at 30 ºC the output voltage is 300 mV, and so on. Power is provided to the sensor chip by connecting its Vs pin to +5 V of Arduino Uno. A small piezoelectric active buzzer is connected to pin 2 of the Arduino Uno at node **ALARM**. Figure 6.17 shows the circuit diagram of the project.

Figure 6.16: LM35DZ temperature sensor chip.

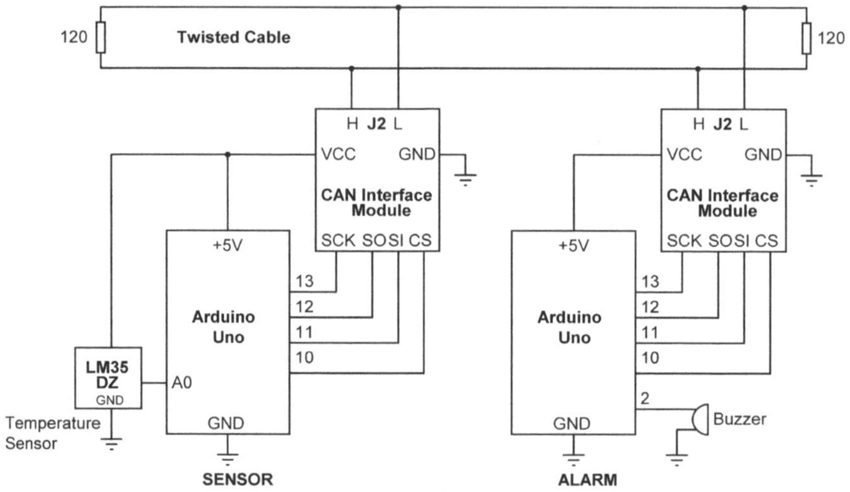

Figure 6.17: Circuit diagram of the project.

Node: SENSOR program listing

Figure 6.18 shows the program listing of node **SENSOR** (program: **SENSOR**). At the beginning of the program, the required header files are included in the program. Inside the **setup()** function, the CAN bus is configured. The ADC converter on Arduino Uno is 10-bits wide, thus having 1024 steps. The ADC reference voltage is 5 V (5000 mV). Therefore, the data read from the ADC must be multiplied by 5000/1024 in order to convert the reading into physical voltage in millivolts. This voltage is then divided by 10 to find the actual measured voltage in degrees Celsius. The temperature is then converted into an integer and loaded into byte 0 of array data and is transmitted over the CAN bus.

```
/*---------------------------------------------------------
                         SENSOR
                         ======

This program reads the ambient temperature and sends it to
node ALARM. If the temperature is above a preset value then
node ALARM activates a buzzer

Author: Dogan Ibrahim
File  : SENSOR
Date  : December 2022
----------------------------------------------------------*/
#include <SPI.h>
#include <mcp2515.h>

#define CS 10
int T = A0;
int val = 0;
float temp;

struct can_frame MyMsg;
MCP2515 mcp2515(CS);                           // SPI CS

void setup()
{
  SPI.begin();
  mcp2515.reset();
  mcp2515.setBitrate(CAN_500KBPS, MCP_8MHZ);  // CAN bus config
  mcp2515.setNormalMode();                     // Normal mode
}

void loop()
{
  val = analogRead(T);
  temp = val * 5000.0 / 1024.0;                // Convert to mV
  temp = temp / 10.0;                          // COnvert to degrees C
  MyMsg.can_id = 0x30;                         // CAN ID
  MyMsg.can_dlc = 8;                           // 8 bytes
  MyMsg.data[0] = (int)temp;                   // Temp in data[0]
  MyMsg.data[1] = 0x00;                        // Remainders 0
  MyMsg.data[2] = 0x00;
  MyMsg.data[3] = 0x00;
  MyMsg.data[4] = 0x00;
```

```
    MyMsg.data[5] = 0x00;
    MyMsg.data[6] = 0x00;
    MyMsg.data[7] = 0x00;

    mcp2515.sendMessage(&MyMsg);                    // Send message
    delay(3000);                                    // Wait 3 secs
}
```

Figure 6.18: Program: ***SENSOR***.

Node: ALARM program listing

Figure 6.19 shows the program listing of node **ALARM** (program: **ALARM**). The temperature preset value is set to 22 degrees Centigrade in this program. If the temperature is greater than this value, then the buzzer is activated to indicate an alarm condition.

```
/*----------------------------------------------------------
                      ALARM
                      =====

This program received the temperature from node SENSOR.If
thetemperature is above a preset value then a buzzer is
activated. The preset value is set to 22 Degrees C here

Author: Dogan Ibrahim
File  : ALARM
Date  : December 2022
----------------------------------------------------------*/
#include <SPI.h>
#include <mcp2515.h>

#define CS 10
int ReceivedMessage;
int Buzzer = 2;                                    // Buzzer at port 2
int PresetTemp = 22;                               // Preset temperature

struct can_frame MyMsg;
MCP2515 mcp2515(CS);

void setup()
{
  SPI.begin();
  mcp2515.reset();
  mcp2515.setBitrate(CAN_500KBPS, MCP_8MHZ);       // Config CAN bus
  mcp2515.setNormalMode();                         // Normal mode
  pinMode(Buzzer, OUTPUT);                         // Buzzer is output
  digitalWrite(Buzzer, 0);                         // Buzzer OFF
```

```
}

void loop()
{
  if(mcp2515.readMessage(&MyMsg) == MCP2515::ERROR_OK)
  {
    ReceivedMessage = MyMsg.data[0];                  // REceived data
    if(ReceivedMessage > PresetTemp)
       digitalWrite(Buzzer, 1);
    else
       digitalWrite(Buzzer, 0);
  }
}
```

Figure 6.19: Program: **ALARM**.

Testing the project: Construct the project as shown in the circuit diagram and power up both Arduino Unos. Increase the temperature of the sensor chip (e.g., by touching the sensor chip or placing something hot near the chip). You should see that the buzzer will sound when the temperature is just above the preset value (you may like to display the temperature on the Serial Monitor to confirm the operation of the project).

6.7 Project 5: Temperature request

Description: This project consists of two nodes, called **TEMPREQ** and **TEMPSENSE**. Node **TEMPREQ** makes requests to node **TEMPSENSE** to get the ambient temperature. Characters **T?** are sent as the request. Node **TEMPSENSE** sends the ambient temperature to node **TEMPREQ** where the temperature is displayed on the Serial Monitor. Temperature is read using an analog temperature sensor chip. The request is made every 5 seconds.

Block Diagram: The block diagram of the project is shown in Figure 6.20.

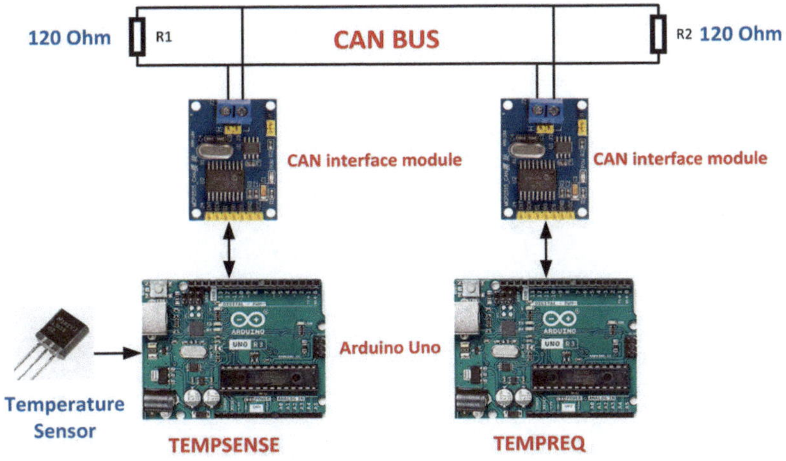

Figure 6.20: Block diagram of the project.

Circuit Diagram: The circuit diagram of the project is shown in Figure 6.21. An LM-35DZ-type temperature sensor chip is connected to port A0 of node **TEMPSENSE** as in the previous project.

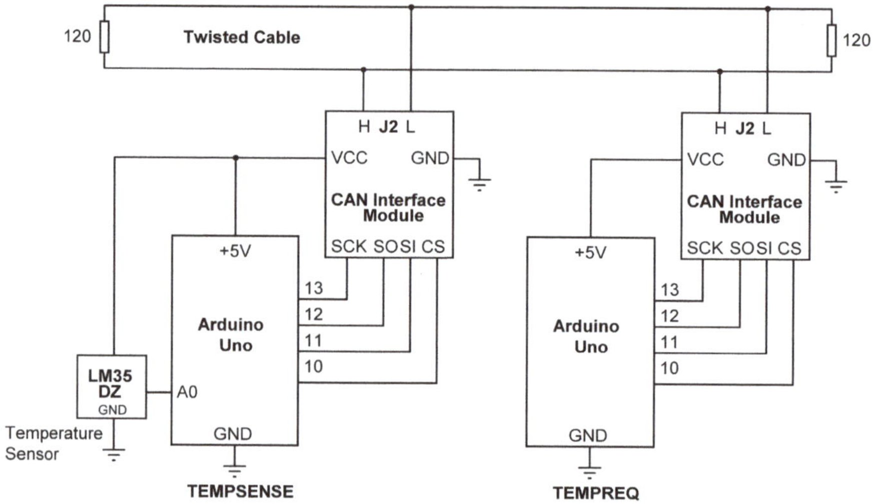

Figure 6.21: Circuit diagram of the project.

Node: TEMPREQ program listing

Figure 6.22 shows the program listing (program **TEMPREQ**). The **setup()** function is similar to the other previous programs, except that here the Serial Monitor is also initialized. Inside the main program loop, requests are sent over the CAN bus to get the ambient temperature. A request is the characters **T?**. Node **TEMPSENSE** upon receiving a request sends the temperature reading to this node where the readings are displayed on the Serial Monitor. Requests are made every 5 seconds.

```
/*----------------------------------------------------------
                      TEMPREQ
                      =======

This program sends requests to node TEMPSENSE to read the
ambient temperature. The received temperature is displayed
on the Serial Monitor. A request is the characters T?.
Requests are made at every 5 seconds

Author: Dogan Ibrahim
File  : TEMPREQ
Date  : December 2022
----------------------------------------------------------*/
#include <SPI.h>
#include <mcp2515.h>
```

```
#define CS 10
int ReceivedMessage;

struct can_frame MyMsg;
MCP2515 mcp2515(CS);

void setup()
{
  SPI.begin();
  mcp2515.reset();
  mcp2515.setBitrate(CAN_500KBPS, MCP_8MHZ);       // Config CAN bus
  mcp2515.setNormalMode();                         // Normal mode
  Serial.begin(9600);                              // Serial Monitor
}

void loop()
{
  MyMsg.can_id = 0x30;                             // CAN ID
  MyMsg.can_dlc = 2;                               // 2 bytes
  MyMsg.data[0] = 'T';                             // Load T
  MyMsg.data[1] = '?';                             // Load ?
  mcp2515.sendMessage(&MyMsg);                     // Send message

  if(mcp2515.readMessage(&MyMsg) == MCP2515::ERROR_OK)
  {
    ReceivedMessage = MyMsg.data[0];               // REceived data
    Serial.print("Temperature = ");
    Serial.println(ReceivedMessage);
    Serial.println("");
  }
  delay(5000);                                     // Wait 5 secs
}
```

Figure 6.22: Program: **TEMPREQ**.

Node: TEMPSENSE program listing

Figure 6.23 shows the program listing (program **TEMPSENSE**). Again, the **setup()** function is similar to other programs. Inside the main program loop, the program waits until message **T?** is received. After receiving this message, the ambient temperature is read from sensor LM35DZ and is sent to node **TEMPREQ** where it is displayed on the Serial Monitor.

```
/*--------------------------------------------------------------
                    TEMPSENSE
                    =========

This program gets requests from node TEMPREQ and then sends
the ambient temperature to node TEMPREQ. A request is the
characters T?

Author: Dogan Ibrahim
File  : TEMPSENSE
Date  : December 2022
---------------------------------------------------------------*/
#include <SPI.h>
#include <mcp2515.h>

#define CS 10
int T = A0;
int val = 0;
float temp;
char msg1, msg2;

struct can_frame MyMsg;
MCP2515 mcp2515(CS);                          // SPI CS

void setup()
{
  SPI.begin();
  mcp2515.reset();
  mcp2515.setBitrate(CAN_500KBPS, MCP_8MHZ);  // CAN bus config
  mcp2515.setNormalMode();                    // Normal mode
}

void loop()
{
  if(mcp2515.readMessage(&MyMsg) == MCP2515::ERROR_OK)
  {
    msg1 = MyMsg.data[0];                     // Received data1
    msg2 = MyMsg.data[1];                     // REceived data2

    if(msg1 == 'T' && msg2 == '?')            // If T? received
    {
      val = analogRead(T);
      temp = val * 5000.0 / 1024.0;           // Convert to mV
      temp = temp / 10.0;                     // COnvert to degrees C
      MyMsg.can_id = 0x30;                    // CAN ID
      MyMsg.can_dlc = 1;                      // 1 byte
```

```
        MyMsg.data[0] = (int)temp;                    // Temp in data[0]
        mcp2515.sendMessage(&MyMsg);                  // Send message
      }
    }
  }
```

Figure 6.23: Program: **TEMPSENSE**.

Testing the project: Construct the project as shown in the circuit diagram and power up both Arduino Unos. Start the Serial Monitor. You should see the temperature displayed every 5 seconds on the Serial Monitor.

6.8 Project 6: Temperature request with a pushbutton

Description: This project is similar to the previous project, but here requests are made by pressing a button. Two nodes are used in this project: **REQBUTTON** and **TEMPSENSE**. Pressing the button **SEND** on node **REQBUTTON** sends characters **T?** to node **TEMPSENSE**, as in the previous project. After receiving a request, the node **TEMPSENSE** sends the temperature reading to node **REQBUTTON** where it is displayed on the Serial Monitor.

Block Diagram: Figure 6.24 shows the block diagram of the project.

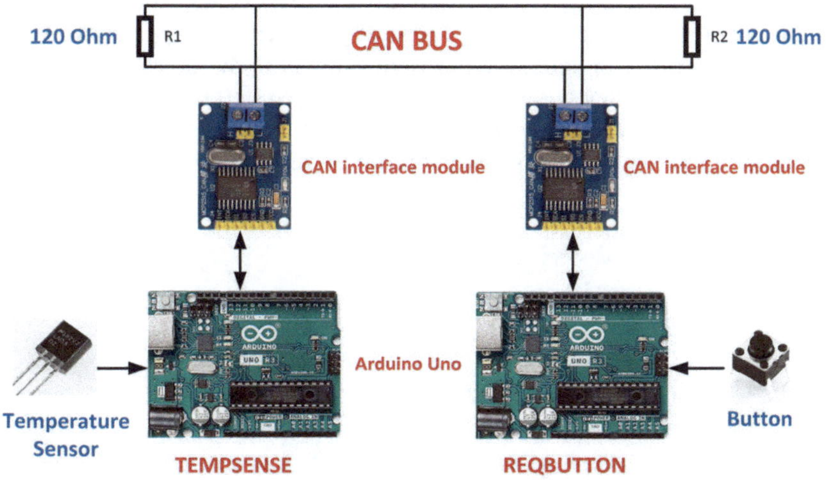

Figure 6.24: Block diagram of the project.

Circuit diagram: The circuit diagram of the project is shown in Figure 6.25. Button **SEND** is connected to port 2 of node **REQBUTTON** and is configured as an input with the internal pull-up resistor enabled. The temperature sensor chip type LM35DZ is connected to port A0 as in the previous project.

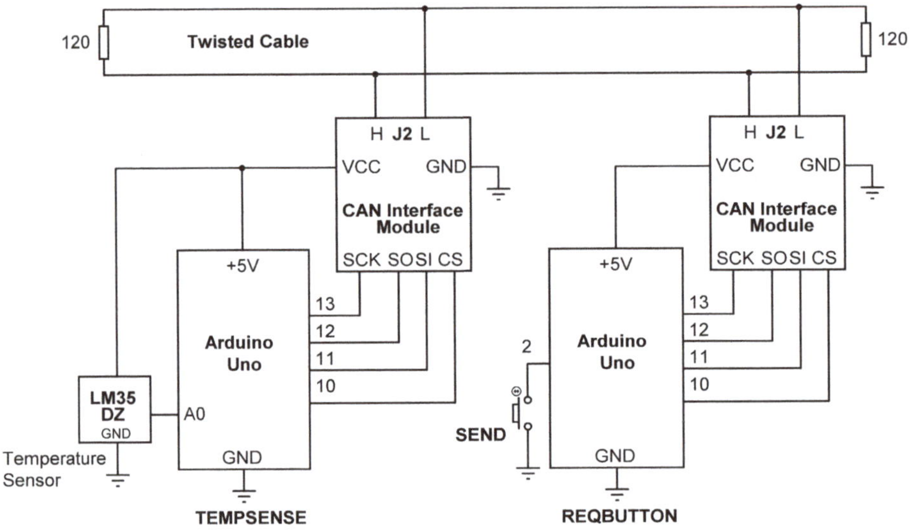

Figure 6.25: Circuit diagram of the project.

Node: REQBUTTON program listing:

Figure 6.26 shows the program listing (program **REQBUTTON**). The **setup()** function is similar to the other previous programs, except that here the Serial Monitor is also initialized. Inside the main program loop, the program waits until the button is pressed and then sends a request over the CAN bus to get the ambient temperature. A request is the characters **T?**. Node **TEMPSENSE** upon receiving a request sends the temperature reading to this node where the readings are displayed on the Serial Monitor.

```
/*-------------------------------------------------------
                      REQBUTTON
                      =========

This program sends requests to node TEMPSENSE when a button
is pressed. TEMPSENSE sends the ambient temperature to this
node which then displays the temperature on Serial Monitor

Author: Dogan Ibrahim
File  : REQBUTTON
Date  : December 2022
----------------------------------------------------------*/
#include <SPI.h>
#include <mcp2515.h>

#define CS 10
int ReceivedMessage;
int Button = 2;                                   // Button port
```

```
struct can_frame MyMsg;
MCP2515 mcp2515(CS);

void setup()
{
  SPI.begin();
  mcp2515.reset();
  mcp2515.setBitrate(CAN_500KBPS, MCP_8MHZ);        // Config CAN bus
  mcp2515.setNormalMode();                          // Normal mode
  pinMode(Button, INPUT_PULLUP);                    // Button is input
  Serial.begin(9600);                               // Serial Monitor
}

void loop()
{
  while(digitalRead(Button) == 1);                  // Wait until button pressed
  while(digitalRead(Button) == 0);                  // Wait until button is
released
  MyMsg.can_id = 0x30;                              // CAN ID
  MyMsg.can_dlc = 2;                                // 2 bytes
  MyMsg.data[0] = 'T';                              // Load T
  MyMsg.data[1] = '?';                              // Load ?
  mcp2515.sendMessage(&MyMsg);                      // Send message

  if(mcp2515.readMessage(&MyMsg) == MCP2515::ERROR_OK)
  {
    ReceivedMessage = MyMsg.data[0];                // REceived data
    Serial.print("Temperature = ");
    Serial.println(ReceivedMessage);
    Serial.println("");
  }
  delay(10);
}
```

Figure 6.26: Program: **REQBUTTON**.

Node: TEMPSENSE program listing:
This program is the same as in Figure 6.23.

Testing the project: Construct the project as shown in the circuit diagram and power up both Arduino Unos. Start the Serial Monitor. Press the button **SEND** to send a request to get and display the temperature on the Serial Monitor.

6.9 Project 7: RGB display with buttons

Description: In this project, there are two nodes: **RGBLED** and **BUTTONS**. **RGB** is an LED module with three LEDs inside emitting red, green, and blue light. The LED has a common pin (e.g., common cathode) that is usually connected to ground through a current-limiting resistor. Node **BUTTONS** has three pushbuttons labeled R, G, and B. Pressing button R sends a command to node **RGBLED** to display Red, pressing G displays green, and so on.

Block diagram: Figure 6.27 shows the block diagram of the project.

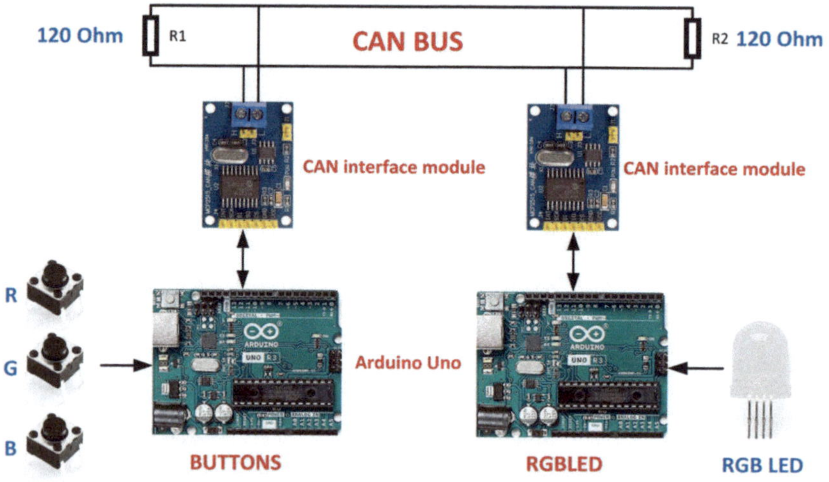

Figure 6.27: Block diagram of the project.

Circuit diagram: The circuit diagram is shown in Figure 6.28. Buttons R, G, and B are connected to port pins 2, 3, and 4 respectively of node **BUTTONS**. The Red, Green, and Blue pins of the RGB LED are connected to port pins 2, 3, and 4 of node **RGBLED** respectively through a current-limiting resistor.

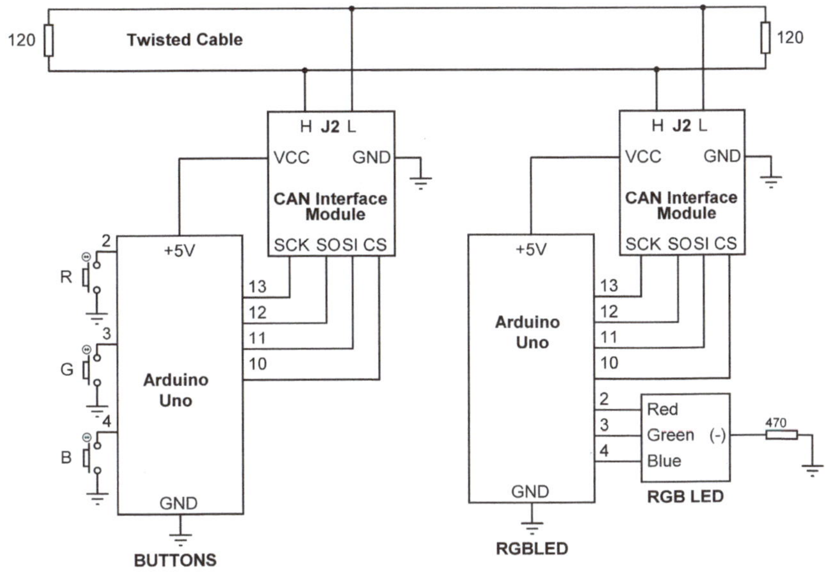

Figure 6.28: Circuit diagram of the project.

Figure 6.29 shows a typical RGB LED. RGB LEDs are of two types: common-cathode and common-anode. In this project, a common-cathode type is used. Note that the long lead is the common cathode pin.

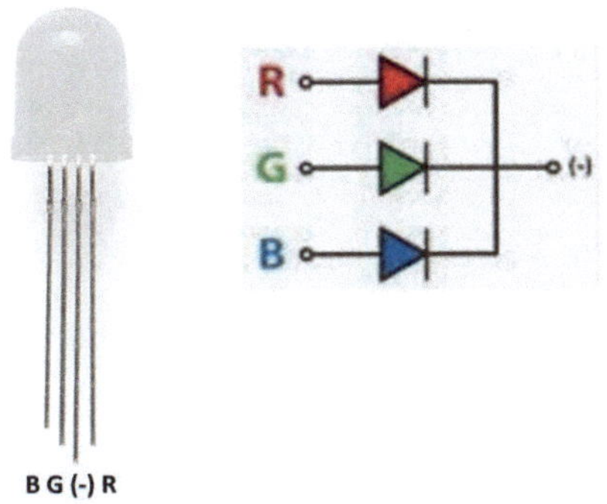

Figure 6.29: Typical RGB LED.

Node: BUTTONS program listing

Figure 6.30 shows the program listing (Program: **BUTTONS**). Inside the **setup()** function, the CAN bus is configured and buttons are configured as inputs with internal pull-up resistors enabled. Inside the main program loop, the state of the buttons is checked and if a button is pressed then a number is sent over the CAN bus to node **RGBLED** to turn ON the required LED. The following numbers sent over the CAN bus correspond to LEDs:

RED LED
GREEN LED
BLUE LED

```
/*---------------------------------------------------------------
                      BUTTONS
                      =======

Three buttons are connected to this node. Pressing a button
turns ON the appropriate LED colour on node RGBLED. For example,
pressing R button turns ON the RED LED. The following data is
sent over the CAN bus:

1   Turn ON RED LED
2   Turn ON GREEN LED
3   Turn ON BLUE LED

Author: Dogan Ibrahim
File  : BUTTONS
Date  : December 2022
-----------------------------------------------------------------*/
#include <SPI.h>
#include <mcp2515.h>

#define CS 10
int ReceivedMessage;
int R = 2;                                     // R button
int G = 3;                                     // G button
int B = 4;                                     // B button
int flag = 0;

struct can_frame MyMsg;
MCP2515 mcp2515(CS);

void setup()
{
  SPI.begin();
  mcp2515.reset();
  mcp2515.setBitrate(CAN_500KBPS, MCP_8MHZ);     // Config CAN bus
  mcp2515.setNormalMode();                       // Normal mode
  pinMode(R, INPUT_PULLUP);                      // R Button is input
  pinMode(G, INPUT_PULLUP);                      // G button is input
  pinMode(B, INPUT_PULLUP);                      // B button is input
}

void loop()
```

```
{
  if(digitalRead(R) == 0)                              // R pressed
  {
    while(digitalRead(R) == 0);
    MyMsg.data[0] = 1;
    flag = 1;
  }

  if(digitalRead(G) == 0)                              // G pressed
  {
    while(digitalRead(G) == 0);
    MyMsg.data[0] = 2;
    flag = 1;
  }

  if(digitalRead(B) == 0)                              // B pressed
  {
    while(digitalRead(B) == 0);
    MyMsg.data[0] = 3;
    flag =1;
  }

  if(flag == 1)                                        // If any button pressed
  {
    flag = 0;
    MyMsg.can_id = 0x30;                               // CAN ID
    MyMsg.can_dlc = 1;                                 // 1 bytes
    mcp2515.sendMessage(&MyMsg);                       // Send message
  }

  delay(20);
}
```

Figure 6.30: Program: **BUTTONS**.

Node: RGBLED program listing

Figure 6.31 shows the program listing (Program: **RGBLED**). Inside the **setup()** function, the 3 LED pins are configured as outputs, and they are turned OFF. Inside the main program loop, the message received from node **BUTTONS** is examined and the required colour of the LED is turned ON.

```
/*------------------------------------------------------------------
                        RGBLED
                        ======

A RGB LED is connected to this node. LED colours Red, Green, or
Blue are lit if 1, 2, or  are received over the CAN bus respectively

Author: Dogan Ibrahim
File  : RGBLED
Date  : December 2022
-------------------------------------------------------------------*/
#include <SPI.h>
#include <mcp2515.h>

#define CS 10
int LED;
int Red = 2;                                          // Red LED
int Green = 3;                                        // Green LED
int Blue = 4;                                         // Blue LED

struct can_frame MyMsg;
MCP2515 mcp2515(CS);

void setup()
{
  SPI.begin();
  mcp2515.reset();
  mcp2515.setBitrate(CAN_500KBPS, MCP_8MHZ);          // Config CAN bus
  mcp2515.setNormalMode();                            // Normal mode
  pinMode(Red, OUTPUT);                               // LED is output
  pinMode(Green, OUTPUT);                             // LED is output
  pinMode(Blue, OUTPUT);                              // LED is output
  digitalWrite(Red, 0);                               // Red OFF OFF
  digitalWrite(Green, 0);                             // Green OFF
  digitalWrite(Blue, 0);                              // Blue OFF
}

void loop()
{
  if(mcp2515.readMessage(&MyMsg) == MCP2515::ERROR_OK)
  {
    LED = MyMsg.data[0];                              // Received data
    if(LED == 1)
    {
      digitalWrite(Red, 1);
      delay(2000);
```

```
        digitalWrite(Red, 0);
      }
      if(LED == 2)
      {
        digitalWrite(Green, 1);
        delay(2000);
        digitalWrite(Green, 0);
      }
      if(LED == 3)
      {
        digitalWrite(Blue, 1);
        delay(2000);
        digitalWrite(Blue, 0);
      }
    }
}
```

Figure 6.31: Program: **RGBLED***.*

Testing the project: Construct the project as shown in the circuit diagram and power up both Arduino Unos. Press a button on node **BUTTONS**. The corresponding colour LED will light up on node **RGBLED**.

6.10 Project 8: Ambient temperature and humidity display on LCD

Description: In this project, there are two nodes called **THSENSOR** and **THDISP**. Node **THSENSOR** measures the ambient temperature and humidity and sends this information to node **THDISP** where it is displayed on an LCD.

Block diagram: Figure 6.32 shows the block diagram of the project.

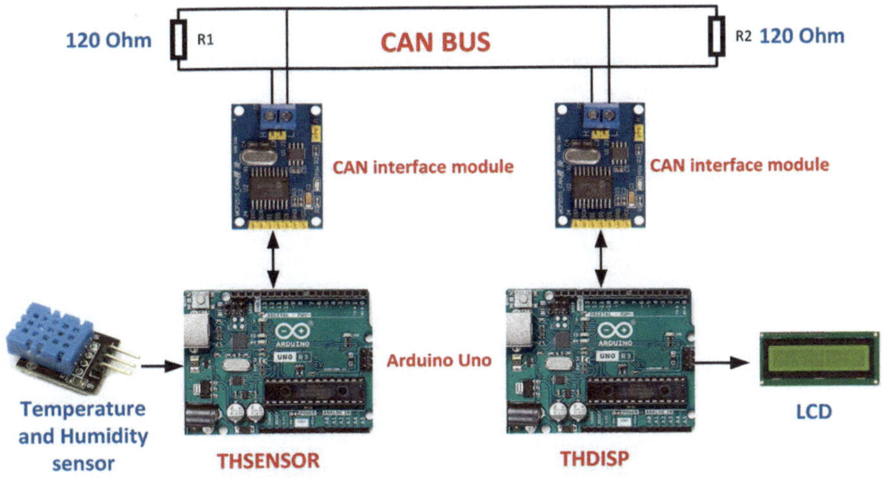

Figure 6.32: Block diagram of the project.

Circuit diagram: The circuit diagram of the project is shown in Figure 6.33. A DHT11-type temperature and humidity sensor is connected to pin 2 at node **THSENSOR**. The temperature and humidity are displayed on an I²C LCD at node **THDISP**.

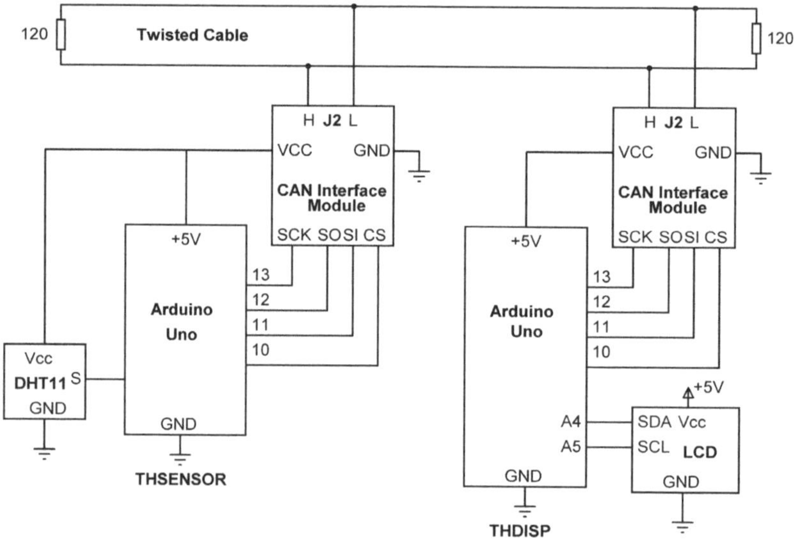

Figure 6.33: Circuit diagram of the project.

DHT11

This is a small sensor that is used to measure temperature and humidity (Figure 6.34). The basic specifications of this sensor are:

- ultra-low cost
- capacitive humidity sensor and thermistor temperature sensor
- 3 to 5 V operation
- 2.5 mA max current use during conversion (while requesting data)
- 20-80% humidity readings with 5% accuracy
- 0-50 °C temperature readings ±2 °C accuracy
- 1 Hz sampling rate
- 15.5 mm × 12 mm × 5.5 mm
- 3 or 4-pin device

Figure 6.34: DHT11 sensor module.

The DHT11 device is usually available as a 3-pin device, and the pin configuration is shown in Figure 6.34.

I²C LCD

The advantage of using an I2C-based LCD is that only two GPIO pins are required to drive the LCD. I²C-based LCD displays are sometimes supplied in two parts: the LCD display, and the I²C controller board. In most standard product distributions, the controller board is soldered to the back of the LCD display as shown in Figure 6.35. The LCD display is basically a 1602-type parallel LCD. The controller board consists of the PCF8574 I²C controller chip (from Texas Instruments or NXP Semiconductors), a small potentiometer to adjust the contrast, I/O interface pins, and address selection jumpers.

Figure 6.35: Standard I2C-based LCD.

The controller board has 4 pins: SCL, SDA, VCC, and GND and these must be soldered to the LCD pins as shown in Figure 6.36 (if it is not already soldered).

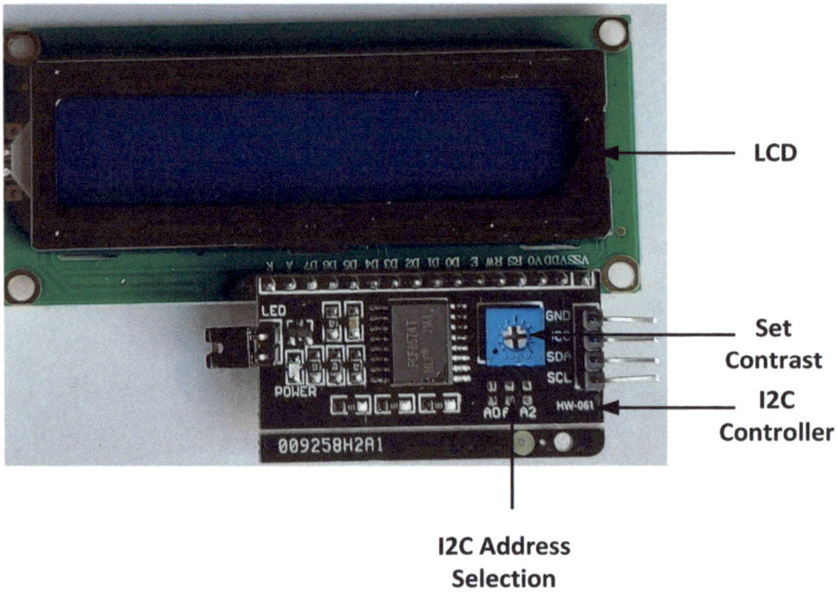

Figure 6.36 Solder the I²C controller board to the LCD.

The I²C address of the PCF8574T chip is selected by 3 jumpers labeled as A0, A1 and A2 on the controller board. By default, the address is set to **0x27** (i.e., no jumper connections).

Arduino Uno I²C pins

The Arduino Uno board has the following I²C pins available:

A4: SDA
A5: SCL

Arduino Uno I²C LCD library

Before using the I²C LCD, you have to add the I²C library to your IDE. Libraries are often distributed as a ZIP file or folder where the name of the folder is the name of the library. Inside the folder, there is a **.cpp** file, an **.h** file, a **keywords.txt** file, an **examples** folder, and other files that may be required by the library. You should not unzip the library.

The steps to add the I²C LCD library are as follows:

- Browse to the following website to locate the library file (the file is also available on the book support webpage):

  https://www.arduinolibraries.info/libraries/liquid-crystal-i2-c

- Download file: **LiquidCrystal_I2C-1.1.2.zip** to a folder on your PC.

- Start the IDE.
- Click **Sketch → Include Library → Add .ZIP Library**.
- Select the saved library file and click **Open**.
- Return to **Sketch → Include Library** menu, and you should see the new library at the bottom of the drop-down menu.
- To test that the library has been added successfully, enter the following lines in a newly created program:

```
#include <Wire.h>
#include <LiquidCrystal_I2C.h>

void setup()
{

}

void loop()
{
}
```

- Compile the program. There should be no errors.

The I^{2}C LCD library supports many functions. Some most commonly used functions are:

init():	initialize LCD (This must be the first function call)
clear():	clear the screen
home():	home the cursor
noBlink():	stop blinking cursor
blink():	enable blinking cursor
noCursor():	hide cursor
cursor():	display cursor
scrollDisplayLeft():	scroll display left
scrollDisplayRight():	scroll display right
noBacklight():	disable backlight
backlight():	enable backlight
setCursor(column, row):	set cursor position (0, 0) is the top left position
print():	print data on LCD
write():	write to LCD

The address of the I^{2}C LCD must be defined at the beginning of the program. For example, if the address is 0x27 and the LCD is 16 columns by 2 rows (i.e., 16×2), then:

LiquidCrystal_I2C lcd(0x27, 16, 2);

Following the above statement, you can call the LCD functions by indexing them with the keyword **lcd**. For example, to initialize the LCD:

```
        lcd.init();                                    // initialize the lcd
```

or, for example, to enable the backlight, use:

```
  lcd.backlight();                         // Enable backlight
```

Arduino Uno DHT11 library

Before using the DHT11, you have to add its library to our IDE. The steps are as follows:

- Copy the library from the following website:
- https://www.electronicwings.com/arduino/dht11-sensor-interfacing-with-arduino-uno
- At the time of writing this book the library file had the name: **Arduino-DHT-master.zip**.
- Start the IDE.
- Click **Sketch → Include Library → Add .ZIP Library**.
- Select the saved library file and click **Open**.
- Return to **Sketch → Include Library** menu, and you should see the new library at the bottom of the drop-down menu.

Node: THSENSOR program listing

Figure 6.37 shows the program listing (Program: **THSENSOR**). At the beginning of the program, the header files of the libraries used in the program are included. Inside the **setup()** function, the CAN bus is configured and the DHT11 pin is set. Inside the main program loop, the ambient temperature and humidity are read using the following functions:

```
   humidity = dht.getHumidity();                 // Get humidity
   temperature = dht.getTemperature();           // Get temperature
```

The temperature and humidity readings are then sent over the CAN bus.

```
/*----------------------------------------------------------
                      THSENSOR
                      ========


This program reads the ambient temperature and humidity from
a DHT11 sensor and sends the data to node THDISP

Author: Dogan Ibrahim
File  : THSENSOR
Date  : December 2022
----------------------------------------------------------*/
#include <SPI.h>
```

```
#include <mcp2515.h>
#include "DHT.h"
DHT dht;
#define CS 10

struct can_frame MyMsg;
MCP2515 mcp2515(CS);                          // SPI CS

int DHTPIN = 2;                               // DHT11 on pin 2
float temperature, humidity;

void setup()
{
  SPI.begin();
  mcp2515.reset();
  mcp2515.setBitrate(CAN_500KBPS, MCP_8MHZ);  // CAN bus config
  mcp2515.setNormalMode();                    // Normal mode
  dht.setup(DHTPIN);                          // Setup DHT11
}

void loop()
{
  humidity = dht.getHumidity();               // GEt humidity
  temperature = dht.getTemperature();         // Get temperature
  MyMsg.can_id = 0x30;                        // CAN ID
  MyMsg.can_dlc = 2;                          // 2 byte
  MyMsg.data[0] = (int)temperature;           // Temp in data[0]
  MyMsg.data[1] = (int)humidity;              // Humidity
  mcp2515.sendMessage(&MyMsg);                // Send message
  delay(5000);                                // Wait 5 seconds
}
```

Figure 6.37: Program: **THSENSOR**.

Node: THDISP program listing

Figure 6.38 shows the program listing (Program: **THDISP**). At the beginning of the program, the header files of used libraries are included, I²C LCD address is set to 0x27, having 16 characters and 2 rows. Inside the setup() function, the CAN bus and the LCD are configured and initialized. Inside the main program loop, the temperature and humidity values are read from the CAN bus and displayed on the LCD. Figure 6.39 shows a typical display.

```
/*--------------------------------------------------------------
                     THDISP
                     ======

This program receives the temperature and humidity from node
THSENSOR and displays on I2C LCD

Author: Dogan Ibrahim
File  : THDISP
Date  : December 2022
--------------------------------------------------------------*/
#include <SPI.h>
#include <mcp2515.h>

#include <Wire.h>
#include <LiquidCrystal_I2C.h>
LiquidCrystal_I2C lcd(0x27, 16, 2);

#define CS 10
int T, H;

struct can_frame MyMsg;
MCP2515 mcp2515(CS);

void setup()
{
  SPI.begin();
  mcp2515.reset();
  mcp2515.setBitrate(CAN_500KBPS, MCP_8MHZ);    // Config CAN bus
  mcp2515.setNormalMode();                      // Normal mode
  lcd.init();                                   // Initialize LCD
  lcd.backlight();                              // Backlight ON
}

void loop()
{
  if(mcp2515.readMessage(&MyMsg) == MCP2515::ERROR_OK)
  {
    T = MyMsg.data[0];                          // Temperature
    H = MyMsg.data[1];                          // Humidity
    lcd.clear();                                // Clear LCD
    lcd.setCursor(0, 0);                        // Cursor at 0,0
    lcd.print(T);                               // Display T
    lcd.print(" C");                            // DIsplay C
    lcd.setCursor(0, 1);                        // Cursor at 0,1
    lcd.print(H);                               // Display H
```

```
    lcd.print(" %");                                       // Display %
  }
}
```

Figure 6.38: Program: **THDISP**.

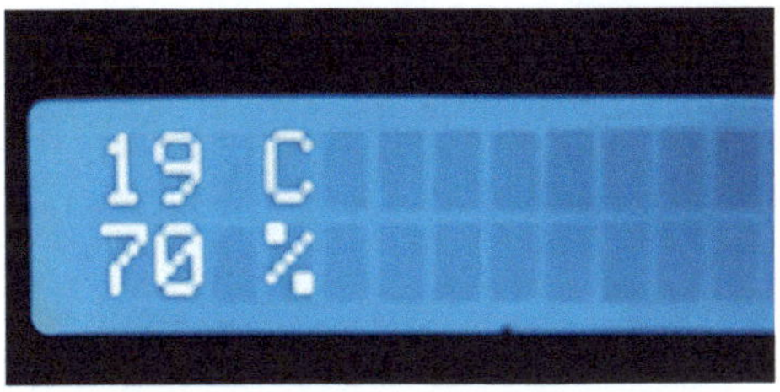

Figure 6.39: Typical display.

Testing the project: Construct the project as shown in the circuit diagram and power up both Arduino Unos. You should see the temperature and humidity displayed on the LCD every 5 seconds.

Modified programs

In the previous project, the temperature and humidity were displayed as integer numbers even though they were read as floating-point numbers. You can display the temperature (and also the humidity) in floating point after converting it into an array of characters and sending it over the CAN bus. Figure 6.40 shows the modified sensor program **THSENSOR2** which sends the temperature as an array of 5 characters. Program **THDISP2** receives the temperature as an array of characters and displays it as a floating-point value with a decimal point.

```
/*---------------------------------------------------------
                       THSENSOR2
                       =========

This program reads the ambient temperature and humidity from
a DHT11 sensor and sends the data to node THDISP.

In this version of the program the temperature is converted
into an array of chars and sent over the CAN bus so that values
after the decimal point can also be displayed

Author: Dogan Ibrahim
File  : THSENSOR2
Date  : December 2022
----------------------------------------------------------*/
#include <SPI.h>
#include <mcp2515.h>
#include "DHT.h"
```

```
DHT dht;
#define CS 10

struct can_frame MyMsg;
MCP2515 mcp2515(CS);                            // SPI CS

int DHTPIN = 2;                                 // DHT11 on pin 2
float temperature, humidity;
char ary[7];

void setup()
{
  SPI.begin();
  mcp2515.reset();
  mcp2515.setBitrate(CAN_500KBPS, MCP_8MHZ);    // CAN bus config
  mcp2515.setNormalMode();                      // Normal mode
  dht.setup(DHTPIN);                            // Setup DHT11
}

void loop()
{
  humidity = dht.getHumidity();                 // GEt humidity
  temperature = dht.getTemperature();           // Get temperature
  MyMsg.can_id = 0x30;                          // CAN ID
  MyMsg.can_dlc = 6;                            // 6 bytes
  dtostrf(temperature, 5, 2, ary);              // Convert to chars
  MyMsg.data[0] = ary[0];                       // Temp in data[0]
  MyMsg.data[1] = ary[1];
  MyMsg.data[2] = ary[2];
  MyMsg.data[3] = ary[3];
  MyMsg.data[4] = ary[4];
  MyMsg.data[5] = (int)humidity;                // Humidity
  mcp2515.sendMessage(&MyMsg);                  // Send message
  delay(5000);                                  // Wait 5 seconds
}
```

Figure 6.40: Program: **THSENSOR2***.*

```
/*-------------------------------------------------------------
                      THDISP2
                      =======

This program receives the temperature and humidity from node
THSENSOR and displays on I2C LCD.

In this version of the program the temperature is displayed
```

```
with values after decimal point

Author: Dogan Ibrahim
File  : THDISP2
Date  : December 2022
---------------------------------------------------------*/
#include <SPI.h>
#include <mcp2515.h>

#include <Wire.h>
#include <LiquidCrystal_I2C.h>
LiquidCrystal_I2C lcd(0x27, 16, 2);

#define CS 10
int H;
char T[7];

struct can_frame MyMsg;
MCP2515 mcp2515(CS);

void setup()
{
  SPI.begin();
  mcp2515.reset();
  mcp2515.setBitrate(CAN_500KBPS, MCP_8MHZ);    // Config CAN bus
  mcp2515.setNormalMode();                      // Normal mode
  lcd.init();                                   // Initialize LCD
  lcd.backlight();                              // Backlight ON
}

void loop()
{
  if(mcp2515.readMessage(&MyMsg) == MCP2515::ERROR_OK)
  {
    for(int i=0; i < 5; i++)T[i] = MyMsg.data[i];
    H = MyMsg.data[5];                          // Humidity
    lcd.clear();                                // Clear LCD
    lcd.setCursor(0, 0);                        // Cursor at 0,0
    lcd.print(T[0]);                            // Display T
    lcd.print(T[1]);
    lcd.print(T[2]);
    lcd.print(T[3]);
    lcd.print(T[4]);
    lcd.print(" C");                            // DIsplay C
    lcd.setCursor(0, 1);                        // Cursor at 0,1
    lcd.print(H);                               // Display H
```

```
        lcd.print(" %");                              // Display %
    }
}
```

Figure 6.41: Program: **THDISP2**.

Figure 6.42 shows an example display.

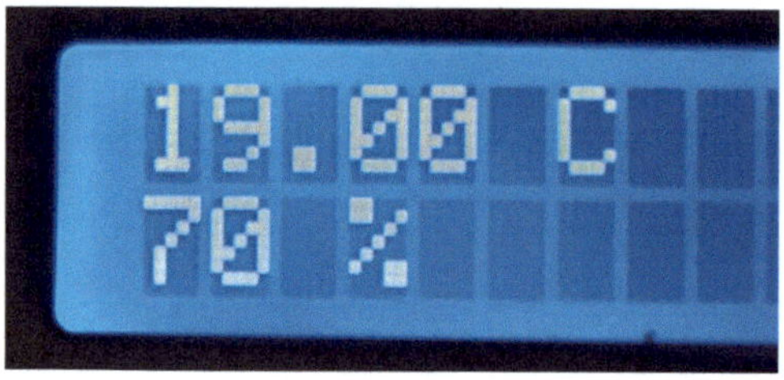

Figure 6.42: Example display.

6.11 Project 9: CAN bus with 3 nodes: External and internal temperature measurement

Description: In this project, 3 nodes are used: **EXTTEMP**, **INTTEMP**, and **DISPTEMP**. Nodes **EXTTEMP** and **INTTEMP** measure the external and internal ambient temperatures. Node **DISPTEMP** receives the temperature data from the other two nodes and displays it on the LCD. The external temperature is displayed every 2 seconds, while the internal temperature is displayed every 5 seconds.

Block diagram: Figure 6.43 shows the block diagram of the project.

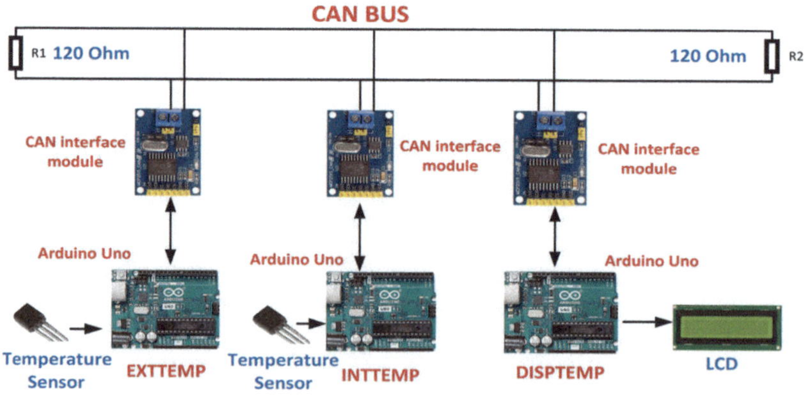

Figure 6.43: Block diagram of the project.

Circuit diagram: The circuit diagram is shown in Figure 6.44. LM35-type analog temperature sensor chips are used in nodes **EXTTEMP** and **INTTEMP**, connected to analog inputs A0 of the Arduino Unos. An I²C LCD is used to display the temperatures on node **DISPTEMP**.

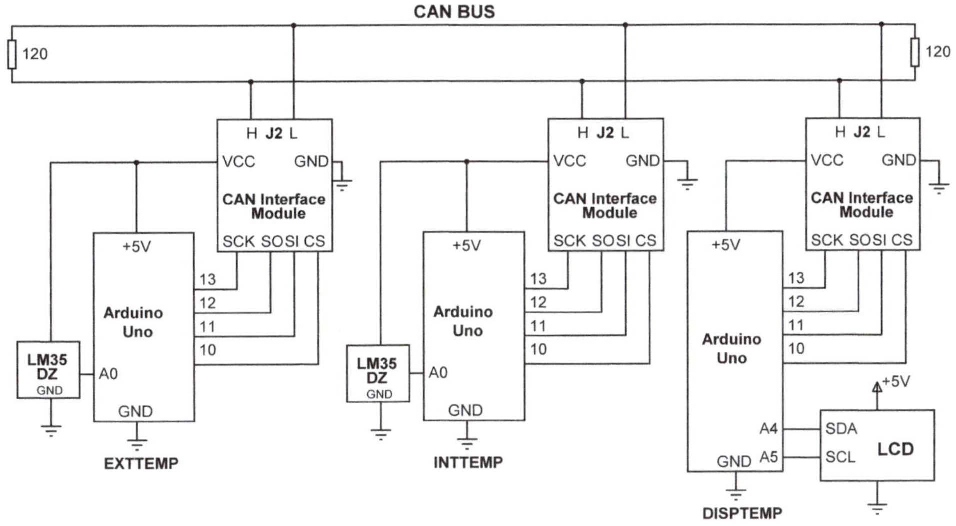

Figure 6.44: Circuit diagram of the project.

Node: EXTTEMP program listing

Figure 6.45 shows the program listing (Program: **EXTTEMP**). In this program, the CAN bus ID is set to 0x30. This program reads the temperature from analog input A0, converts into degrees Celsius, and then sends it over the CAN bus with the CAN ID set to 0x30. Only one byte is sent.

```
/*-----------------------------------------------------------
                    EXTTEMP
                    =======

This program reads the external ambient temperature and sends
it over the CAN bus every 2 seconds

Author: Dogan Ibrahim
File  : EXTTEMP
Date  : December 2022
-----------------------------------------------------------*/
#include <SPI.h>
#include <mcp2515.h>

#define CS 10
int T = A0;
int val = 0;
float temp;

struct can_frame MyMsg;
MCP2515 mcp2515(CS);                          // SPI CS
```

```
void setup()
{
  SPI.begin();
  mcp2515.reset();
  mcp2515.setBitrate(CAN_500KBPS, MCP_8MHZ);  // CAN bus config
  mcp2515.setNormalMode();                    // Normal mode
}

void loop()
{
  val = analogRead(T);
  temp = val * 5000.0 / 1024.0;               // Convert to mV
  temp = temp / 10.0;                         // COnvert to degrees C
  MyMsg.can_id = 0x30;                        // CAN ID
  MyMsg.can_dlc = 1;                          // 1 byte
  MyMsg.data[0] = (int)temp;                  // Temp in data[0]
  mcp2515.sendMessage(&MyMsg);                // Send message
  delay(2000);                                // Wait 2 secs
}
```

Figure 6.45: Program: ***EXTTEMP****.*

Node: INTTEMP program listing

Figure 6.46 shows the program listing (Program: **INTTEMP**). This program is similar to Figure 6.45. The only difference here is that the CAN bus ID is set to 0x31.

```
/*----------------------------------------------------------
                       INTTEMP
                       =======

This program reads the internal ambient temperature and sends
it over the CAN bus every 5 seconds

Author: Dogan Ibrahim
File  : INTTEMP
Date  : December 2022
-----------------------------------------------------------*/
#include <SPI.h>
#include <mcp2515.h>

#define CS 10
int T = A0;
int val = 0;
float temp;
```

```
struct can_frame MyMsg;
MCP2515 mcp2515(CS);                             // SPI CS

void setup()
{
  SPI.begin();
  mcp2515.reset();
  mcp2515.setBitrate(CAN_500KBPS, MCP_8MHZ);   // CAN bus config
  mcp2515.setNormalMode();                     // Normal mode
}

void loop()
{
  val = analogRead(T);
  temp = val * 5000.0 / 1024.0;                // Convert to mV
  temp = temp / 10.0;                          // COnvert to degrees C
  MyMsg.can_id = 0x31;                         // CAN ID
  MyMsg.can_dlc = 1;                           // 1 byte
  MyMsg.data[0] = (int)temp;                   // Temp in data[0]
  mcp2515.sendMessage(&MyMsg);                 // Send message
  delay(5000);                                 // Wait 5 secs
}
```

Figure 6.46: Program: ***INTTEMP****.*

Node: DISPTEMP program listing

Figure 6.47 shows the program listing. At the beginning of the program, the header files of the libraries used are included and LCD address is set to 0x27, and it is configured for 16 characters, 2 rows. Inside the **setup()** function, the CAN bus is configured, and the LCD is initialized. Inside the main program loop, the program checks the CAN ID of the received messages. If the ID is 0x30, then the message has come from the node **EXTTEMP** and therefore the temperature reading is displayed at the top row of the LCD. If, on the other hand, the ID is 0x31, then the message has come from the node **INTTEMP** and therefore the temperature reading is displayed at the bottom row of the LCD. This process is continued until stopped by the user.

```
/*----------------------------------------------------------
                      DISPTEMP
                      ========

This program receives the external and internal temperature
readings over the CAN bus and displays them on the LCD. The
external temperature is displayed on the top row while the
internal temperature is displayed on the bottom row

Author: Dogan Ibrahim
```

```
File  : DISPTEMP
Date  : December 2022
-----------------------------------------------------------*/
#include <SPI.h>
#include <mcp2515.h>

#include <Wire.h>
#include <LiquidCrystal_I2C.h>
LiquidCrystal_I2C lcd(0x27, 16, 2);

#define CS 10
int T, id;

struct can_frame MyMsg;
MCP2515 mcp2515(CS);

void setup()
{
  SPI.begin();
  mcp2515.reset();
  mcp2515.setBitrate(CAN_500KBPS, MCP_8MHZ);    // Config CAN bus
  mcp2515.setNormalMode();                      // Normal mode
  lcd.init();                                   // Initialize LCD
  lcd.backlight();                              // Backlight ON
}

void loop()
{
  if(mcp2515.readMessage(&MyMsg) == MCP2515::ERROR_OK)
  {
    T = MyMsg.data[0];                          // Temperature
    id = MyMsg.can_id;                          // CAN ID
    if(id == 0x30)                              // If External
    {
      lcd.setCursor(0, 0);                      // Cursor at 0,0
      lcd.print("EXT: ");                       // Display EXT
      lcd.print(T);                             // Display T
      lcd.print(" C");                          // Display C
    }

    if(id == 0x31)                              // If Internal
    {
      lcd.setCursor(0, 1);                      // Cursor at 0,1
      lcd.print("INT: ");                       // Display INT
      lcd.print(T);                             // Display T/
      lcd.print(" C");                          // Display C
```

```
        }
        delay(20);
    }
}
```

Figure 6.47: Program: **DISPTEMP***.*

Testing the project

Construct the circuits and apply power to all 3 nodes. You should see the external and internal temperatures displayed on the LCD. The external temperature is displayed at the top row, while the internal temperature is displayed at the bottom row of the LCD.

Figure 6.48 shows a typical display.

Figure 6.48: Typical display.

6.12 Using acceptance masks and filters

In a CAN bus system with many nodes, you may want to receive messages from certain nodes only. This is done using the acceptance masks and acceptance filters on receive nodes. A 1 in a mask bit accepts the ID bit whether it is 0 or 1. In the previous project, you wanted to receive messages from node IDs 0x30 and 0x31. As described in Chapter 5, you can set the masks and filters as follows (assuming standard 11-bit IDs, masks and filters):

0x30:	000 0011 0000
0x31:	000 0011 0001
Required Mask:	111 1111 1110 or 0x7FE

You should therefore set the mask to 0x7E and the filter to 0x30 or 0x31.

Figure 6.49 shows the display program (**DISPTEMP2**) of Project 9 where messages from only 0x30 and 0x31 are accepted by the program.

The program code to set MASK0 and Filter 0 is as follows. Notice that **false** in the statements mean that you are using standard 11-bit IDs, masks and filters.

```
mcp2515.setConfigMode();
mcp2515.setFilterMask(MCP2515::MASK0, false, 0x7FE);
mcp2515.setFilter(MCP2515::RXF0, false, 0x30);
mcp2515.setNormalMode();
```

```
/*-----------------------------------------------------------
                      DISPTEMP2
                      =========

This program receives the external and internal temperature
readings over the CAN bus and displays them on the LCD. The
external temperature is displayed on the top row while the
internal temperature is displayed on the bottom row.

In this version of the program acceptance mask and acceptance
filter are used

Author: Dogan Ibrahim
File  : DISPTEMP2
Date  : December 2022
-----------------------------------------------------------*/
#include <SPI.h>
#include <mcp2515.h>

#include <Wire.h>
#include <LiquidCrystal_I2C.h>
LiquidCrystal_I2C lcd(0x27, 16, 2);

#define CS 10
int T, id;

struct can_frame MyMsg;
MCP2515 mcp2515(CS);

void setup()
{
  SPI.begin();
  mcp2515.reset();
  mcp2515.setBitrate(CAN_500KBPS, MCP_8MHZ);     // Config CAN bus
  lcd.init();                                    // Initialize LCD
  lcd.backlight();                               // Backlight ON

  mcp2515.setConfigMode();
  mcp2515.setFilterMask(MCP2515::MASK0, false, 0x7FE);
  mcp2515.setFilter(MCP2515::RXF0, false, 0x30);
  mcp2515.setNormalMode();                       // Normal mode
}

void loop()
{
  if(mcp2515.readMessage(&MyMsg) == MCP2515::ERROR_OK)
```

```
    {
      T = MyMsg.data[0];                              // Temperature
      id = MyMsg.can_id;                              // CAN ID
      if(id == 0x30)                                  // If External
      {
        lcd.setCursor(0, 0);                          // Cursor at 0,0
        lcd.print("EXT: ");                           // Display EXT
        lcd.print(T);                                 // Display T
        lcd.print(" C");                              // Display C
      }

      if(id == 0x31)                                  // If Internal
      {
        lcd.setCursor(0, 1);                          // Cursor at 0,1
        lcd.print("INT: ");                           // Display INT
        lcd.print(T);                                 // Display T/
        lcd.print(" C");                              // Display C
      }
      delay(20);
    }
}
```

Figure 6.49: Program: **DISPTEMP2**.

6.13 Project 10: CAN bus with 3 nodes: External and internal temperature measurement with pushbuttons

Description: This project is similar to the previous project. Additionally, here, two pushbuttons are used to request the external or the internal temperature to be displayed. Three nodes are used in the project: **TEMPX**, **TEMPIN**, **TEMPXINDISP**. The pushbuttons, called **EXT** and **INT** are connected to node **TEMPXINDISP**. Pressing the button **EXT** gets the external temperature from node **TEMPX** and displays it on the LCD. Similarly, pressing the button **INT** gets the internal temperature from node **TEMPIN** and displays it on LCD.

Block diagram: Figure 6.50 shows the block diagram of the project.

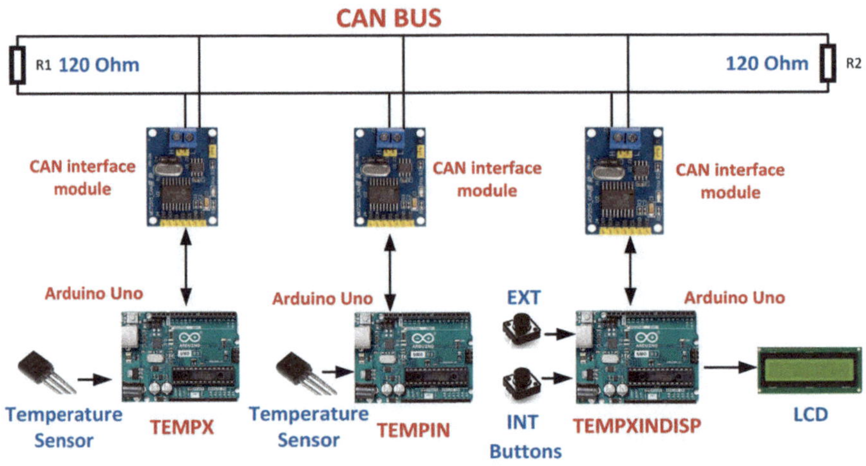

Figure 6.50: Block diagram of the project.

Circuit diagram: The circuit diagram is shown in Figure 6.51. It is similar to Figure 6.44, but here additionally 2 buttons are connected to ports 2 and 3 of node **TEMPXINDISP**.

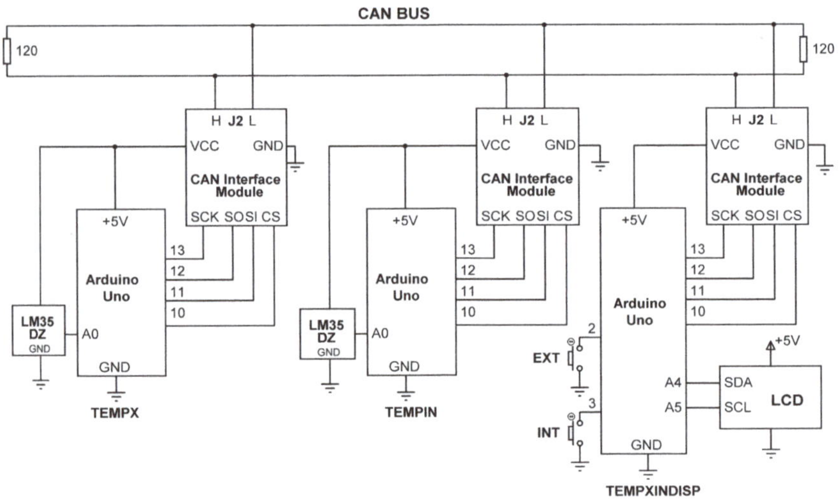

Figure 6.51: Circuit diagram of the project.

PDL of the project: Figure 6.52 shows the operation of the project as a PDL (Program Description Language). PDL (sometimes called pseudocode) is the description of the operation of a computer program in English-like statements.

```
TEMPXINDISP
BEGIN
    Configure CAN bus and initialize LCD
    DO FOREVER
        IF Button EXT is pressed THEN
            Send 1 over CAN bus with ID = 0x41
        ENDIF
        IF Button INT is pressed THEN
            Send 2 over CAN bus with ID = 0x42
        ENDIF
        IF message is received over CAN bus THEN
            Get message ID
            IF ID = 0x30 THEN
                Display external temperature
            ENDIF
            IF ID = 0x31 THEN
                Display internal temperature
            ENDIF
        ENDIF
    ENDDO

TEMPX
BEGIN
    Configure CAN bus
    DO FOREVER
        IF message received over CAN bus THEN
            IF ID = 1 THEN
                Send temperature reading with ID = 0x30
            ENDIF
        ENDIF
    ENDDO

TEMPIN
BEGIN
    Configure CAN bus
    DO FOREVER
        IF message received over CAN bus THEN
            IF ID = 2 THEN
                Send temperature reading with ID = 0x31
            ENDIF
        ENDIF
    ENDDO
```

Figure 6.52: PDL of the project.

Node: TEMPX program listing

Figure 6.53 shows the program listing (Program: **TEMPX**). Inside the **setup()** function, the CAN bus is configured. Inside the main program loop, the program polls to check if there is data over the CAN bus. If the data is 1, then a request for external temperature reading is made. External temperature is sent over the bus as an integer value with the CAN bus id of 0x30.

```
/*----------------------------------------------------------
                      TEMPX
                      =====

This program reads the external ambient temperature when a
request is received over the CAN bus. The reading is sent
with message of 1 and ID = 0x30

Author: Dogan Ibrahim
File  : TEMPX
Date  : December 2022
-----------------------------------------------------------*/
#include <SPI.h>
#include <mcp2515.h>

#define CS 10
int T = A0;
int val = 0;
float temp;
int msg;

struct can_frame MyMsg;
MCP2515 mcp2515(CS);                                // SPI CS

void setup()
{
  SPI.begin();
  mcp2515.reset();
  mcp2515.setBitrate(CAN_500KBPS, MCP_8MHZ);        // CAN bus config
  mcp2515.setNormalMode();                          // Normal mode
}

void loop()
{
  if(mcp2515.readMessage(&MyMsg) == MCP2515::ERROR_OK)
  {
    msg = MyMsg.data[0];                            // Temperature
    val = analogRead(T);
    temp = val * 5000.0 / 1024.0;                   // Convert to mV
```

```
    temp = temp / 10.0;                              // COnvert to degrees C

    if(msg == 1)                                     // Request for external
    {
      MyMsg.can_id = 0x30;                           // CAN ID
      MyMsg.can_dlc = 1;                             // 1 byte
      MyMsg.data[0] = (int)temp;                     // Temp in data[0]
      mcp2515.sendMessage(&MyMsg);                   // Send message
    }
  }
}
```

Figure 6.53: Program: **TEMPX***.*

Node: TEMPIN program listing

Figure 6.54 shows the program listing (Program: **TEMPIN**). Inside the **setup()** function, the CAN bus is configured. Inside the main program loop, the program polls to check if there is data over the CAN bus. If the data is 2, then a request for internal temperature reading is made. Internal temperature is sent over the bus as an integer value with the CAN bus id of 0x31.

```
/*----------------------------------------------------------
                        TEMPIN
                        ======

This program reads the internal ambient temperature when a
request is received over the CAN bus. The reading is sent
with message of 1 and ID = 0x31

Author: Dogan Ibrahim
File  : TEMPIN
Date  : December 2022
-----------------------------------------------------------*/
#include <SPI.h>
#include <mcp2515.h>

#define CS 10
int T = A0;
int val = 0;
float temp;
int msg;

struct can_frame MyMsg;
MCP2515 mcp2515(CS);                            // SPI CS

void setup()
```

```
{
  SPI.begin();
  mcp2515.reset();
  mcp2515.setBitrate(CAN_500KBPS, MCP_8MHZ);        // CAN bus config
  mcp2515.setNormalMode();                          // Normal mode
}

void loop()
{
  if(mcp2515.readMessage(&MyMsg) == MCP2515::ERROR_OK)
  {
    msg = MyMsg.data[0];                            // Temperature
    val = analogRead(T);
    temp = val * 5000.0 / 1024.0;                   // Convert to mV
    temp = temp / 10.0;                             // COnvert to degrees C

    if(msg == 2)                                    // Request for internal
    {
      MyMsg.can_id = 0x31;                          // CAN ID
      MyMsg.can_dlc = 1;                            // 1 byte
      MyMsg.data[0] = (int)temp;                    // Temp in data[0]
      mcp2515.sendMessage(&MyMsg);                  // Send message
    }
  }
}
```

Figure 6.54: Program: ***TEMPIN****.*

Node: TEMPXINDISP program listing

Figure 6.55 shows the program listing (Program: **TEMPXINDISP**). At the beginning of the program, headers of the used libraries are included and LCD is configured with address 0x27, 16 characters and 2 rows. Inside the **setup()** function, the CAN bus is configured and the two buttons **EXT** and **INT** are configured as inputs with the internal pull-up resistors enabled. The program then checks if a button is pressed. If the button **EXT** is pressed, then 1 is sent over the CAN bus. If, on the other hand, the button **INT** is pressed, then 2 is sent over the bus. The program then reads the external or the internal temperature values from nodes **TEMPX** or **TEMPIN** and displays them on the LCD. The temperature readings are displayed at the top row of the LCD in the following format:

EXT: nn C or INT: nn C

```
/*----------------------------------------------------------------
                      TEMPXINDISP
                      ===========

Two buttons are connected to this node. Pressing button EXT
```

```
sends a request over the CAN bus to get the external temperature
from node EXTTEMP. The temperature is then displayed on the LCD.
Similarly, pressing button INT sends a request over the CAN bus
to get the internal temperature from node INTTEMP and then displays
this temperature on the LCD.

Author: Dogan Ibrahim
File  : TEMPXINDISP
Date  : December 2022
-----------------------------------------------------------------*/
#include <SPI.h>
#include <mcp2515.h>
#include <Wire.h>
#include <LiquidCrystal_I2C.h>
LiquidCrystal_I2C lcd(0x27, 16, 2);

#define CS 10
int ReceivedMessage;
int EXT = 2;                                          // R button
int INT = 3;                                          // G button
int T, ID, flag = 0;

struct can_frame MyMsg;
MCP2515 mcp2515(CS);

void setup()
{
  SPI.begin();
  mcp2515.reset();
  mcp2515.setBitrate(CAN_500KBPS, MCP_8MHZ);          // Config CAN bus
  mcp2515.setNormalMode();                            // Normal mode
  pinMode(EXT, INPUT_PULLUP);                         // EXT Button is input
  pinMode(INT, INPUT_PULLUP);                         // INT button is input
  lcd.init();                                         // Initialize LCD
  lcd.backlight();                                    // Backlight ON
}

void loop()
{
  if(digitalRead(EXT) == 0)                           // EXT pressed
  {
    while(digitalRead(EXT) == 0);
    MyMsg.data[0] = 1;
    flag = 1;
  }
```

```
    if(digitalRead(INT) == 0)                           // INT pressed
    {
      while(digitalRead(INT) == 0);
      MyMsg.data[0] = 2;
      flag = 1;
    }

    if(flag == 1)                                       // If any button pressed
    {
      flag = 0;
      MyMsg.can_id = 0x41;                              // CAN ID
      MyMsg.can_dlc = 1;                                // 1 bytes
      mcp2515.sendMessage(&MyMsg);                      // Send message
    }

    if(mcp2515.readMessage(&MyMsg) == MCP2515::ERROR_OK)
    {
      ID = MyMsg.can_id;
      T = MyMsg.data[0];
      lcd.clear();
      lcd.setCursor(0, 0);

      if(ID == 0x30)                                  // If External
      {
        lcd.print("EXT: ");                           // Display EXT
        lcd.print(T);                                 // Display T
        lcd.print(" C");                              // Display C
      }

      if(ID == 0x31)                                  // If Internal
      {
        lcd.print("INT: ");                           // Display INT
        lcd.print(T);                                 // Display T/
        lcd.print(" C");                              // Display C
      }
    }
  }
```

Figure 6.55: Program: ***TEMPXINDISP***.

Testing the project: Construct the circuits and apply power to all 3 nodes. Press the button **EXT**, you should see the external temperature displayed on the LCD. Press the button **INT**, and you should this time see the internal temperature displayed.

Chapter 7 • Error Conditions on the CAN Bus

7.1 Overview

Before going into the details of CAN bus error types, it is worthwhile to look at the "bit stuffing" mechanisms on the bus.

7.2 Bit stuffing

The CAN standard specifies that any bits of the same polarity (recessive or dominant) on the bus that are longer than 5 bits long is a violation of the standard. In fact, this standard was used to send error frames on the bus, consisting of 6 dominant bits in sequence.

In some applications, it may be a requirement to send more than 5 bits of the same polarity (e.g., the data bits may contain more than 5 bits of the same polarity). This type of situation is handled on the bus by the transmitting node inserting a bit of opposite polarity after the 5th bit. The receiving node then removes this bit. This mechanism called bit stuffing allows synchronizing the transmit and receive operations to prevent timing errors. Note that error and overload frames are transmitted without bit stuffing. Also, during a reception, if the 6th bit is the same as the 5th, then a Stuffing Error occurs on the bus.

Bit stuffing is allowed from the SOF (Start of Frame) field to the CRC (Cyclic Redundancy Check) field (see Figure 7.1). However, Bit stuffing is not allowed in the static fields of a frame. i.e., it is not allowed in the following fields:

- CRC delimiter
- ACK field
- EOF field
- Interframe gap

Figure 7.2 shows an example of bit stuffing, where the transmitting node added a recessive bit after the 5th dominant bit. The receiving node removed this bit, and thus more than 5 dominant bits have successfully been transmitted on the bus.

Similarly, in Figure 7.3, the transmitting node added a dominant bit after the 5th recessive bit. The receiving node again removed the stuffed bit, and thus more than 5 recessive bits have successfully been transmitted on the bus.

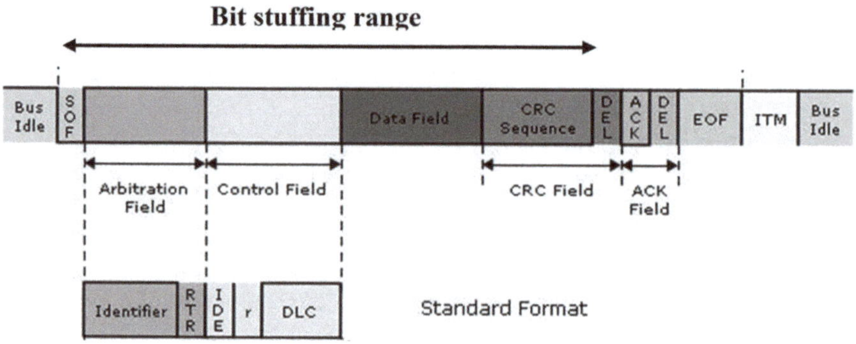

Figure 7.1: Fields where bit stuffing can be used.

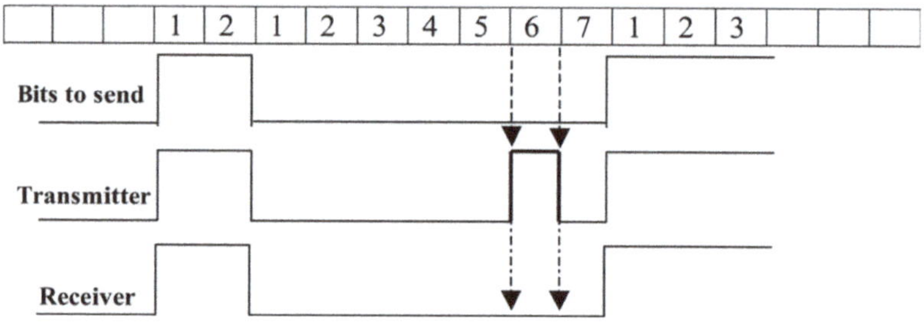

Figure 7.2: Bit stuffing example (adding a recessive bit).

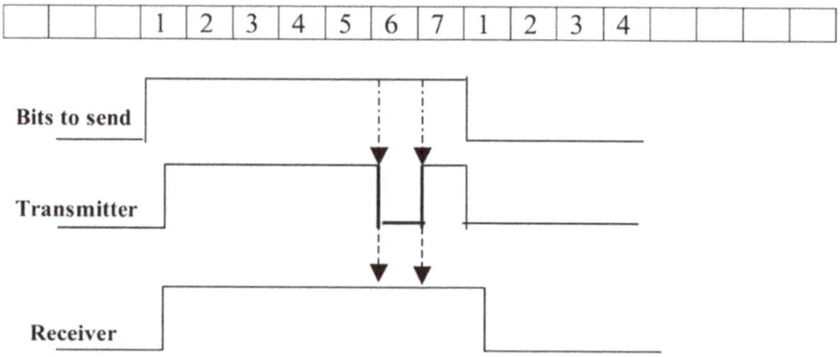

Figure 7.3: Bit stuffing example (adding a dominant bit).

7.3 CAN Bus error detection

There are five types of errors that can occur on the bus. If any of these errors are observed, the error frame is transmitted, which causes the current frame on the bus to be declared as faulty and discarded by all nodes on the bus. The error types are:

- Bit error
- Bit stuffing error
- CRC error
- Frame error
- ACK error

Now, let's have a look at these error conditions in more detail.

7.3.1 Bit error

When a node transmits a bit on the bus, it also monitors the bus and compares the transmitted bit with the actual level on the bus. A bit error is said to happen when the transmitted bit is not the same as the bit level on the bus.

Note that a bit error will not happen during the arbitration phase, where a node transmits a recessive bit while another node transmits a dominant bit. Also, while a node is transmitting an error frame with 6 consecutive recessive bits if a dominant bit is detected on the bus this will not create a bit error.

7.3.2 Bit stuffing error

A bit stuffing error will occur if the 6th bit on the bus is the same as the 5th. Note that during transmission of an error or overload frame, no bit stuffing errors will occur (even though 6 consecutive dominant bits are sent).

7.3.3 CRC error

The transmitting node calculates the CRC using all the bits from the SOF to the end of data field, and then inserts the calculated value into the frame just after the data field. Any receiving nodes also calculate the CRC and expect to find the same value as the received value inside the frame. A CRC error will happen if the CRC value sent by the transmitting node (inside the frame) is not the same as the CRC value calculated by any receiving node (calculated using the received bits from SOF to the end of the data field).

When a CRC error is observed, the error frame is sent after a delay of 2-bit times so that there is no confusion with the ACK slot.

7.3.4 Frame error

Frame errors relate to errors in the format of a frame. A frame error will be observed when the static field of a frame is not as expected. For example, if the ACK delimiter or the CRC delimiter is missing or if the EOF field contains dominant bits, or if the interframe gap contains dominant bits.

7.3.5 ACK Error

The transmitting node monitors the bus and expects a dominant bit during the ACK slot. An ACK error will happen if the transmitting node detects a recessive bit during the ACK slot.

7.4 CAN bus fault confinement

The CAN bus is marked by high reliability. It is capable of detecting errors, as well as correcting itself if any abnormalities are detected. Each node has an error signaling ability which depends on the historic behavior of that node. For example, if a node keeps generating errors continuously, then that node can voluntarily remove itself from the bus so that it does not cause a deadlock on the bus.

There are three fundamental states that define the error signaling state of each node:

- Error Active
- Error Passive
- Bus Off

The normal state of a node is the Error Active state. When a node is in Error Active state, it can send all frames, including error frames.

When a node is in Error Passive state, the node can send all frames except the error frame. i.e., the node cannot participate in error determination on the bus.

A node is isolated from the bus and stops communicating when it is in Bus Off state.

Two internal error counts are maintained by the CAN controller hardware of each node in order to determine the state of the node at any time. These counters are:

- Transmit Error Counter (TEC)
- Receive Error Counter (REC)

The TEC counter increments whenever an error is detected within the node while sending a frame. Also, the TEC counter is decremented whenever a frame is successfully sent. In a similar way, the REC counter is incremented if an error occurs while receiving a frame and is decremented whenever a successful receive operation is performed.

If any of the two counters in a node become greater than 127, the node goes into the Error Passive mode. In this mode, the node can still send and receive frames, but cannot destroy frames on the bus (i.e., it cannot send error frames). Because the error counters get decremented, it is possible that a node, which is in Error Passive mode, can return to the normal Error Active mode if both of its counters are equal to or less than 127.

If on the other hand, the error counters increment further, the node stays in Error Passive mode until the TEC counter becomes greater than 255. At this point, the node moves to the Bus Off mode. In this mode, the node shuts down and stops sending or receiving frames on the bus. A node that is in Bus Off mode stays in this mode and self-recovery is not possible. Re-initializing the controller or re-initializing the overall bus system should move the node back to the normal Error Active mode. Figure 7.4 shows the CAN bus error states.

The CAN bus was developed such that one of the most stringent requirements set by automotive applications get fulfilled by its standards. It is so reliable that, based on calculation, if a network based on 250 kbps operates for 5 hours a day for a year at an average bus load of 25%, an undetected error occurs only once per 1000 years!

In the above example, an undetected error means that bits in a message get corrupted in such a way that the CRC algorithm does not detect them, and a **faulty** message frame is transmitted on the bus. What this means in automotive terms is that, if a car is driven for

5 hours a day, every day for a year, an error will go undetected in the car electronics once every 1000 years, which is well above the lifetime of the car!

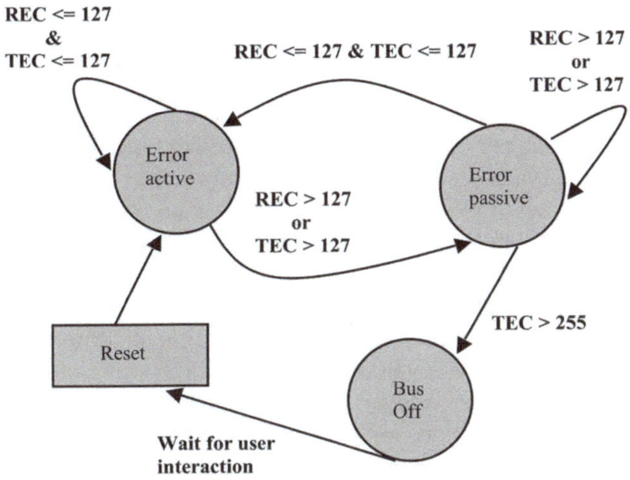

Figure 7.4: CAN bus error states.

7.5 Summary

The CAN bus is a highly reliable bus structure. This chapter has described all error conditions that can occur on the CAN bus. In addition, fault confinement methods have been explained. It is shown that a node that is reporting an error continuously can voluntarily disconnect itself from the bus, thus making the bus available to other nodes in the system.

Chapter 8 • CAN Bus Analyzers

8.1 Overview

CAN bus analyzers are hardware and software tools that can be useful during the development of CAN bus-based projects. This chapter presents a number of these tools briefly.

8.2 CAN Bus analyzers

Essentially, a CAN bus analyzer consists of a small piece of hardware (called analyzer hardware) and dedicated software (called analyzer software, which usually runs on a PC). One end of the analyzer hardware is attached to a CAN bus as a node, while the other end is usually connected to a PC via USB. Once the analyzer software is activated, the analyzer hardware starts to collect all the frames sent over the bus, with time stamping. The collected data can be analyzed, and any data transmission or timing errors on the bus can be detected. A bus analyzer can be very useful during the development of a CAN bus-based project. In addition, these tools are valuable teaching aids where students can easily analyze the collected data structures and timing details and learn about CAN bus data communication principles. In this section, we shall be looking at some of the more popular CAN bus analyzer tools commercially available.

8.2.1 Microchip, Inc. CAN bus analyzer

This is a low-cost tool (see Figure 8.1) developed by Microchip Inc that can be used during the development and debugging of a CAN network. The tool supports CAN 2.0B and ISO11898-2 and includes graphical software that makes it easier to analyze a CAN bus network. The basic features of this tool are:

- Supports CAN 2.0B and ISO11898-2 standards
- PC user interface for functions such as configuration, trace, transmit, filter, etc.
- Direct access to CAN H and CAN L, CAN TX and CAN RX signals for debugging
- Flexible CAN bus interface options such as standard DB9 connector or screw terminals
- Software control of termination resistance and LED display for status, traffic, and bus errors

Figure 8.1: Microchip, Inc. CAN bus analyzer (www.microchip.com).

A sample display from this tool is shown in Figure 8.2.

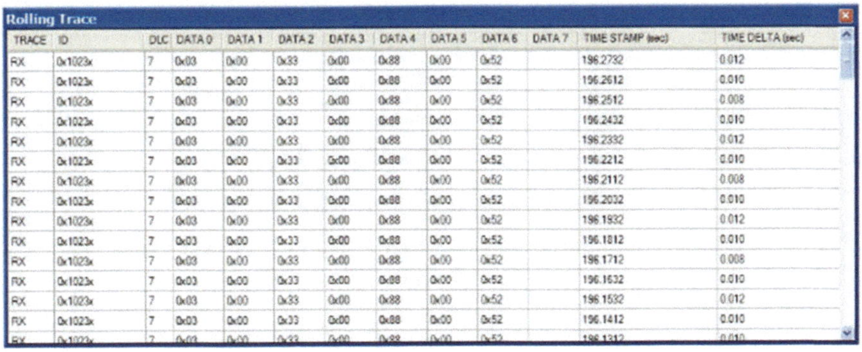

Rolling Trace

TRACE	ID	DLC	DATA 0	DATA 1	DATA 2	DATA 3	DATA 4	DATA 5	DATA 6	DATA 7	TIME STAMP (sec)	TIME DELTA (sec)
RX	0x1023x	7	0x03	0x00	0x33	0x00	0x88	0x00	0x52		196.2732	0.012
RX	0x1023x	7	0x03	0x00	0x33	0x00	0x88	0x00	0x52		196.2612	0.010
RX	0x1023x	7	0x03	0x00	0x33	0x00	0x88	0x00	0x52		196.2512	0.008
RX	0x1023x	7	0x03	0x00	0x33	0x00	0x88	0x00	0x52		196.2432	0.010
RX	0x1023x	7	0x03	0x00	0x33	0x00	0x88	0x00	0x52		196.2332	0.012
RX	0x1023x	7	0x03	0x00	0x33	0x00	0x88	0x00	0x52		196.2212	0.010
RX	0x1023x	7	0x03	0x00	0x33	0x00	0x88	0x00	0x52		196.2112	0.008
RX	0x1023x	7	0x03	0x00	0x33	0x00	0x88	0x00	0x52		196.2032	0.010
RX	0x1023x	7	0x03	0x00	0x33	0x00	0x88	0x00	0x52		196.1932	0.012
RX	0x1023x	7	0x03	0x00	0x33	0x00	0x88	0x00	0x52		196.1812	0.010
RX	0x1023x	7	0x03	0x00	0x33	0x00	0x88	0x00	0x52		196.1712	0.008
RX	0x1023x	7	0x03	0x00	0x33	0x00	0x88	0x00	0x52		196.1632	0.010
RX	0x1023x	7	0x03	0x00	0x33	0x00	0x88	0x00	0x52		196.1532	0.012
RX	0x1023x	7	0x03	0x00	0x33	0x00	0x88	0x00	0x52		196.1412	0.010
RX	0x1023x	7	0x03	0x00	0x33	0x00	0x88	0x00	0x52		196.1312	0.010

Figure 8.2: Microchip, Inc. CAN bus analyzer sample display.

8.2.2 CANdo

This is a USB-to-CAN bus interface (see Figure 8.3) that gets connected directly between a PC and a CAN bus network. Together with the CANdo application, it provides a powerful means of analyzing any CAN network. This tool supports both CAN 2.0A and CAN 2.0B.

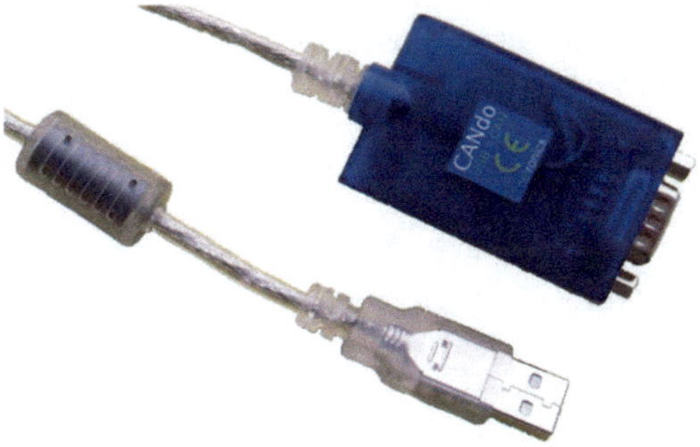

Figure 8.3: CANdo analyzer (www.cananalyser.com).

A sample display from this tool is shown in Figure 8.4.

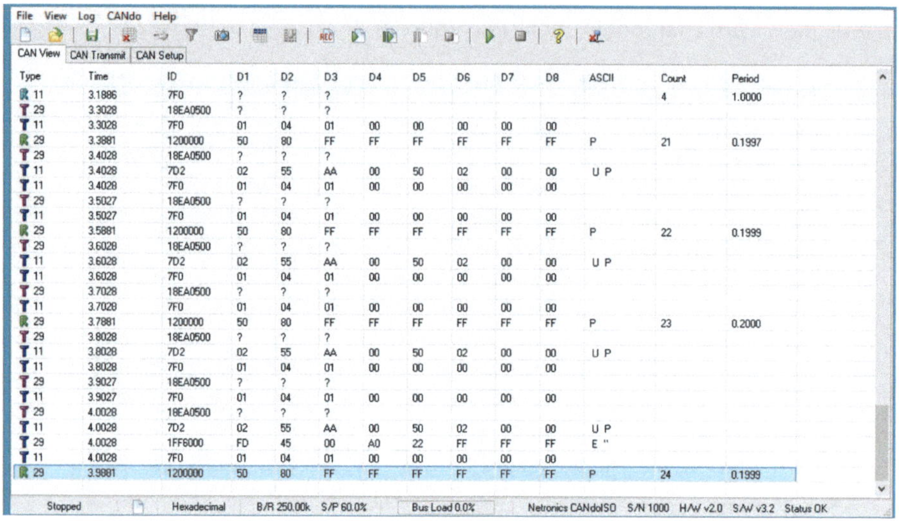

Figure 8.4: Sample display from CANdo.

8.2.3 PCAN-Explorer

The PCAN®-Explorer program from PEAK-System Technik GmbH Darmstadt (Figure 8.5) displays CAN bus activity and can also be used for controlling and interacting with CAN bus systems, for debugging the system, or for a PC program driving it.

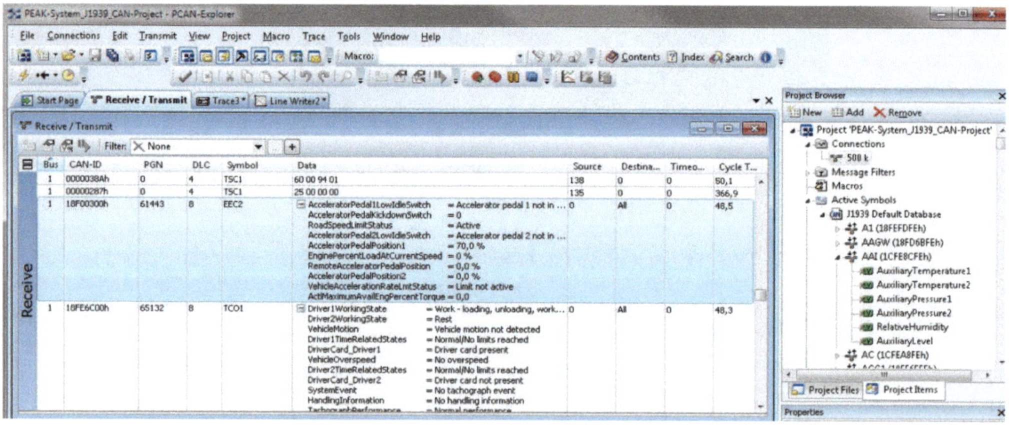

Figure 8.5: PCAN-Explorer. (www.peak-system.com).

Some of the specifications of PCAN-Explorer are:

- Shows all messages received in a receiving list containing message ID, length and data bytes.
- Indication is given of received remote-frames, number and receiving interval.
- Any number of messages can be put into a transmit list to be sent at fixed intervals, manually, from function keys, or as the answer to a remote frame request.

- Errors on the CAN bus are indicated.
- Each CAN message ID can be given a unique name which is then used in place of the ID when messages are logged, making complex systems easy to debug.
- An extensive VBScript Macro language can be used to test or simulate CAN drivers.
- Add-ins allow the users to automate tasks in PCAN Explorer by adding commands to perform these tasks, by adding toolbar buttons to carry out these commands, and by responding to PCAN Explorer commands.

Some of the specifications of the data logger are:

- variable buffer size
- errors can be logged
- logged data can be saved in Excel format so that the data can be analyzed offline using Excel
- message types to be logged can be selected by the user

A data logging example display is shown in Figure 8.6.

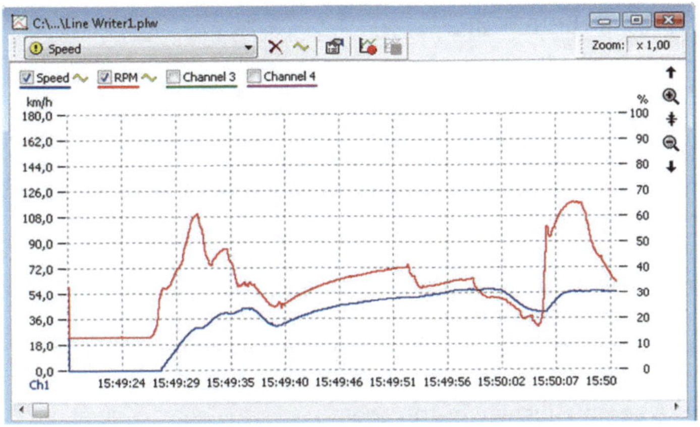

Figure 8.6: Data logging example.

8.2.4 CAN-Bus-Tester 2 (CBT2)

This tool (see Figure 8.7) supports CAN, CANopen, DeviceNet and SAE J1939 systems and is used to analyze, monitor, and troubleshoot a CAN network. The tester provides an overview of the signal-to-noise ratios on the bus, thus enabling causes of failures to be identified and possibly rectified.

Figure 8.7: CAN-Bus-Tester 2 (www.ixxat.com).

Some of the features of this tool are:

- integrated oscilloscope with data interpretation
- comprehensive trigger conditions for fault localization
- monitoring of bus status, bus load, and errors
- wiring test on the bus, including termination resistance, short circuits, loop resistance, line length, etc.
- automatic baud rate detection
- operation using Windows program on a PC
- generation of test protocols
- monitoring reception and monitoring of messages on the bus

Figure 8.8 shows a sample display of this tool.

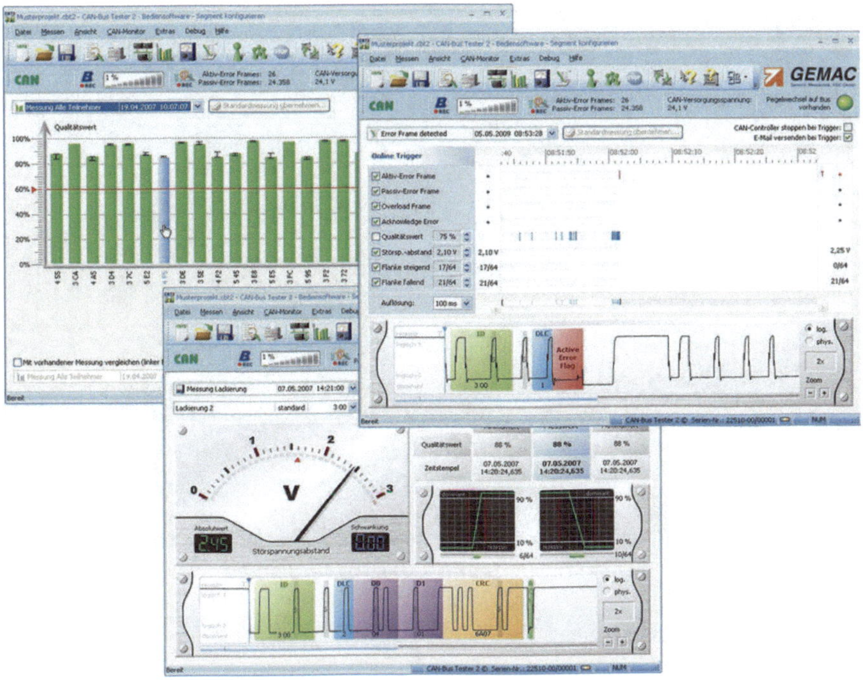

Figure 8.8: Sample display from CBT2.

8.2.5 BitScope Logic

BitScope Logic is a 12-channel logic and protocol analyzer that can be used as a general-purpose logic analyzer, or to analyze protocols such as SPI, I²C, and CAN bus. The system consists of several models of hardware (see Figure 8.9 for Model 10) and software (see Figure 8.10).

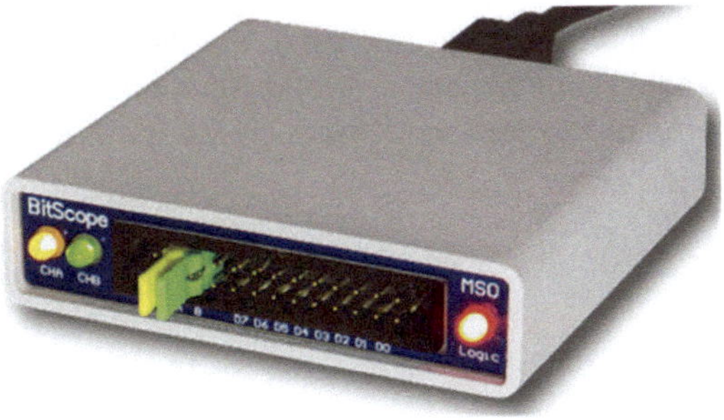

Figure 8.9: BitScope model 10 (www.bitscope.com).

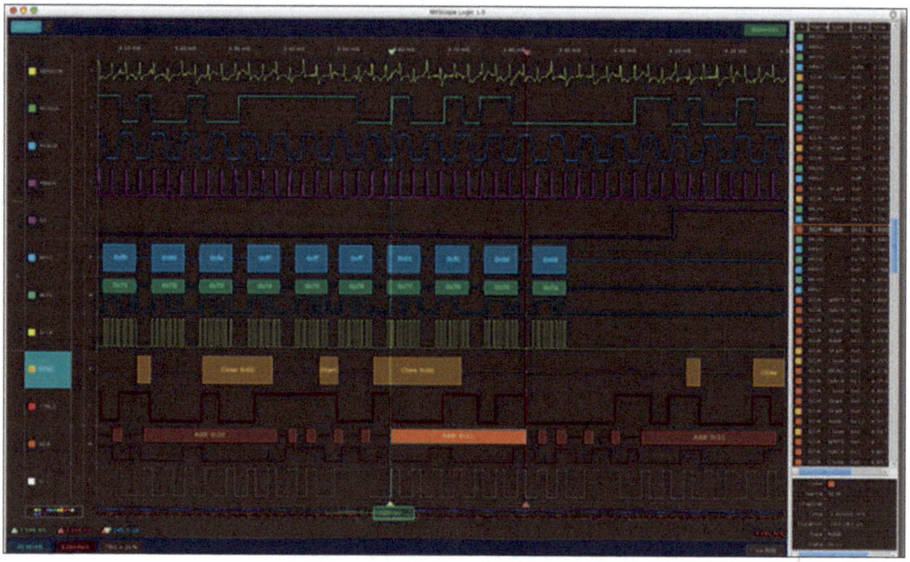

Figure 8.10: Sample BitScope display.

The basic features of BitScope are:

- SPI, CAN, I²C and async serial protocol analysis
- Up to 8 channels
- Packet data inspection
- Built-in (4-channel) oscilloscope
- Color-coded channels with user labels
- Trace navigator, cursors and packet lists

8.2.6 LAP-C logic analyzer

LAP-C is a general-purpose logic analyzer (Figure 8.11) that has a large number of protocol analysis functions. Depending on memory size and functionality, there are several models of this logic analyzer. Some of the protocols that can be analyzed with the LAP-C are:

- SPI
- I²C
- CAN
- RS232/RS485
- LIN
- USB
- 7-segment LED
- PS/2
- Microwire
- 1-wire
- UART

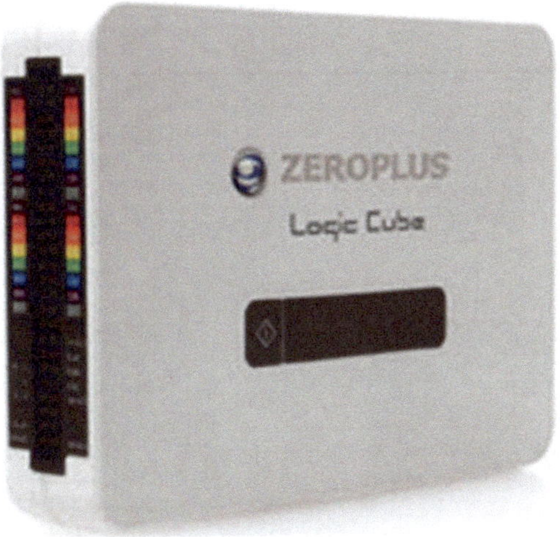

Figure 8.11: LAP-C logic analyzer (www.zeroplus.com).

8.3 Project 11: CAN bus sniffer

Description: In some applications, we may want to know what is on the CAN bus. In this project, we will display the following over the CAN bus. The data will be displayed on the Serial Monitor of the IDE:

Node ID
Size of the message
Message itself

In this project, two Arduino Unos and two CAN bus interface modules are used. The nodes are labelled as: **SND** and **CATCH**. Node **SND** transmits data (random numbers between 1 and 100) over the bus every 5 seconds. Node **CATCH** catches all the data on the bus and displays the message ID, message size, and the message itself.

It is assumed that the bus sniffer (Node: **CATCH**) uses the same bit rate as the other nodes on the bus (in some applications, the bit rate on the bus may not be known).

Block diagram: Figure 8.12 shows the block diagram. You just use a node to catch and display the data on the bus.

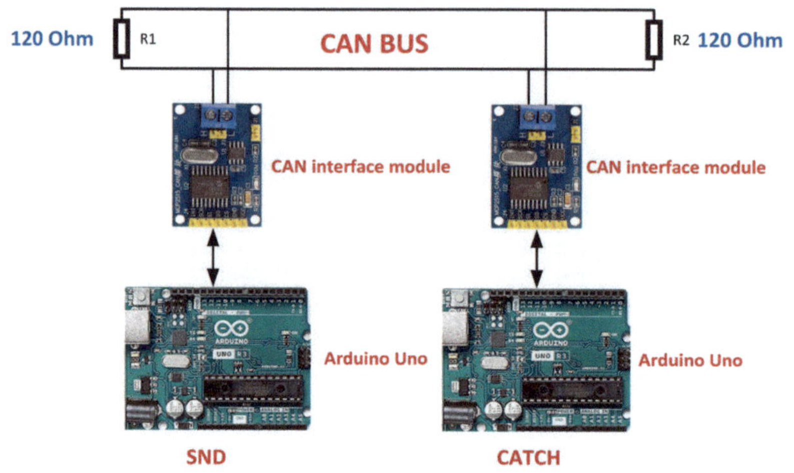

Figure 8.12: Block diagram of the project.

Circuit diagram: The circuit diagram is shown in Figure 8.13. There are no external sensors or components in this project.

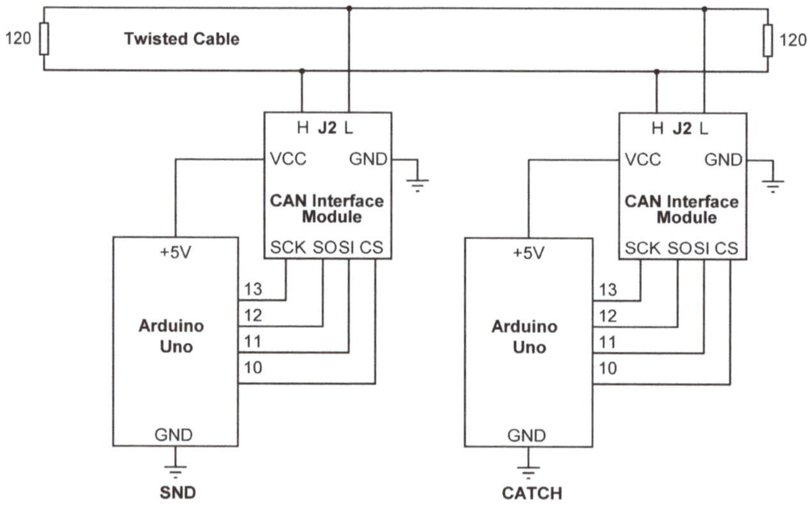

Figure 8.13: Circuit diagram of the project.

Node: SND program listing:

Figure 8.14 shows the program listing (Program: **SND**). At the beginning of the program, the header files *SPI.h* and *mcp2515.h* are included in the program. Inside the **setup()** function, the CAN bus is configured. Inside the main program loop, random numbers are generated between 1 and 100 and these numbers are transmitted over the CAN bus. The byte count is set to 1.

```
/*--------------------------------------------------------------
                      SND
                      ===

This program generates random numbers between 1 and 100 and
sends them over the CAN bus. Node CATCH catches all the
data on the bus and tabulates the message ID, message size,
and the message itself.

Author: Dogan Ibrahim
File  : SND
Date  : December 2022
--------------------------------------------------------------*/
#include <SPI.h>
#include <mcp2515.h>

#define CS 10
int RandomNumber;

struct can_frame MyMsg;
MCP2515 mcp2515(CS);                            // SPI CS

void setup()
{
  SPI.begin();
  mcp2515.reset();
  mcp2515.setBitrate(CAN_500KBPS, MCP_8MHZ);  // CAN bus config
  mcp2515.setNormalMode();                      // Normal mode
}

void loop()
{
  RandomNumber = random(1, 101);                // Random number
  MyMsg.can_id = 0x30;                          // CAN ID
  MyMsg.can_dlc = 1;                            // 1 byte
  MyMsg.data[0] = RandomNumber;                 // Number in data[0]
  mcp2515.sendMessage(&MyMsg);                  // Send message
  delay(5000);                                  // Wait 5 secs
}
```

Figure 8.14: Program: ***SND****.*

Node: CATCH program listing

Figure 8.15 shows the program listing (Program: CATCH). At the beginning of the program, the header files SPI.h and mcp2515.h are included in the program. Inside the **setup()** function, the CAN bus is configured, and Serial Monitor is set to 9600 Baud. Inside the main program loop, the message ID, message size and message content are stored in variables **id**, **msg**, and **sz**, respectively. These variables are displayed in tabular form on the Serial Monitor.

```
/*---------------------------------------------------------
                        CATCH
                        =====

This program receives data over the CAN bus and displays the
message ID, message size, and the contents of the message.
The data is displayed on the Serial Monitor

Author: Dogan Ibrahim
File  : CATCH
Date  : December 2022
----------------------------------------------------------*/
#include <SPI.h>
#include <mcp2515.h>

#define CS 10
int msg, id, sz;
struct can_frame MyMsg;
MCP2515 mcp2515(CS);                              // SPI CS

void setup()
{
  Serial.begin(9600);                             // Serial Monitor
  Serial.println("ID\tSize\tMessage");
  Serial.println("==\t====\t=======\t");

  SPI.begin();
  mcp2515.reset();
  mcp2515.setBitrate(CAN_500KBPS, MCP_8MHZ);      // CAN config
  mcp2515.setNormalMode();                        // Normal mode
}

void loop()
{
  if(mcp2515.readMessage(&MyMsg) == MCP2515::ERROR_OK)
  {
    msg = MyMsg.data[0];
    id = MyMsg.can_id;
    sz = MyMsg.can_dlc;

    Serial.print(id);
    Serial.print("\t");
```

```
    Serial.print(sz);
    Serial.print("\t");
    Serial.println(msg);
  }
}
```

Figure 8.15: Program: **CATCH**.

Testing the project: Connect the circuit as in Figure 8.13 and power up the Arduino Unos. Start the Serial Monitor on node **CATCH**. You should see the data on the CAN bus displayed in a tabular form. An example display is shown in Figure 8.16.

Notice that only 1 byte of data is transmitted in this example for simplicity. But the data size can be extended as required. For example, a **for** loop can be set up with **k** as the loop count and the received data bytes **MyMsg.data[k]** can be displayed as **k** runs from 0 to **MyMsg.can_dlc**.

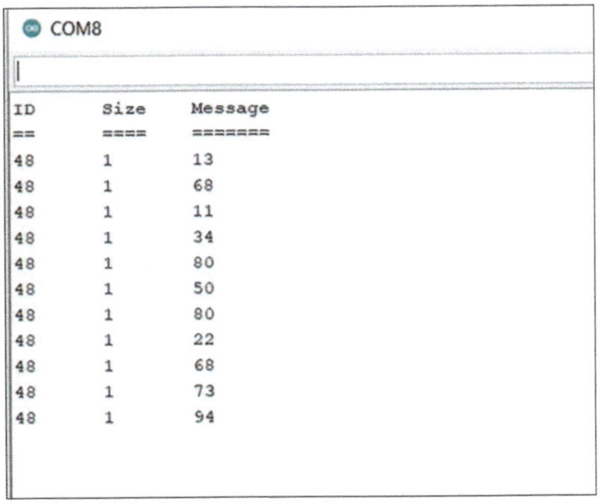

Figure 8.16: Data on the CAN bus.

8.4 Summary

This chapter has described the important topic of CAN bus development tools. Without a development tool, it is very hard to develop a project, especially a complex project based on the CAN bus.

The hardware tools consist of hardware development boards. It is shown that some companies provide integrated development tools, where a complete CAN bus system is offered with a few nodes. Users can simply create a CAN bus by connecting the supplied hardware together. In addition, high level language compilers are also provided so that users can develop, test, and debug their programs easily.

CAN bus analyzers are an important part of developing CAN bus-based projects. With the aid of these analyzers, users can monitor and log the data moving over the CAN bus. For example, during the development of a CAN bus project, programmers can examine the packets on the bus and make sure that the correct data is transferred at the correct times. In addition, CAN bus analyzers can be used as teaching aids to teach the structure of packets and the timing of data packets on a CAN bus. With the aid of the analyzers, data can be logged and then analyzed in detail offline.

Chapter 9 • Raspberry Pi CAN Bus Projects

9.1 Overview

In this chapter, you will be developing projects using the Raspberry Pi development boards connected over a CAN bus through CAN bus interface modules. You will be using the Raspberry Pi 4 in the projects.

9.2 Project 11: Simple Raspberry Pi – Arduino Uno CAN bus communication

Description: In this project, two nodes are used: **RPISND** and **ARCATCH**. Node **RPISND** is the Raspberry Pi node. This node transmits numbers and letters over the CAN bus. **ARCATCH** is the Arduino Uno node that receives these numbers and letters and displays them on the Serial Monitor.

Block diagram: Figure 9.1 shows the block diagram.

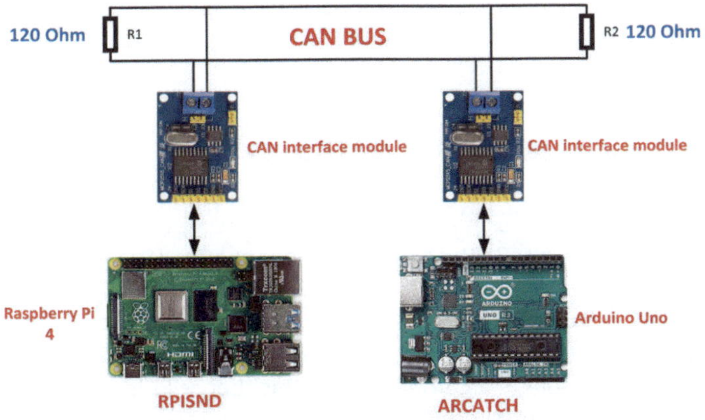

Figure 9.1: Block diagram of the project.

SPI bus pins of Raspberry Pi 4: Figure 9.2 shows the Raspberry Pi 4 pin diagram. There are 2 SPI ports (SPI0 and SPI1) with the following GPIO pins:

SPI0 GPIO pins	SPI function
GPIO 7	CE1
GPIO 8	CE0
GPIO 9	MISO
GPIO 10	MOSI
GPIO 11	SCLK

SPI1 GPIO pins	SPI function
GPIO 16	CE2
GPIO 17	CE1
GPIO 18	CE0
GPIO 19	MISO

GPIO 20 MOSI
GPIO 21 SCLK

By default, SPI0 is enabled. SPI1 can be enabled by entering the following statement in file: */boot/config.txt:*

```
dtoverlay=spi1-3cs
```

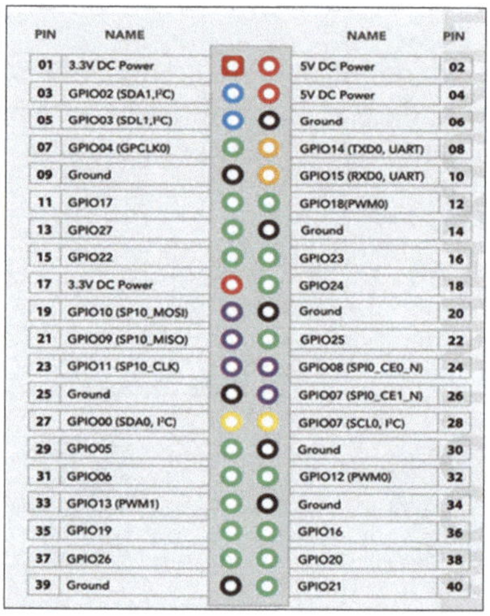

Figure 9.2: Raspberry Pi 4-pin configuration.

Circuit diagram: MCP2515 chip operates with +3.3 to +5 V. Unfortunately, the TJA1050 chip only operates at +5 V and the power supply of the MCP2515 and TJA1050 chips on the CAN bus interface module are common. The Raspberry Pi is not +5 V compatible, so you cannot connect the CAN bus interface bus module directly to Raspberry Pi. Solutions are available though.

1. You can use resistive potential divider circuits to lower the interface module SO and INT pin voltages from 5 V to 3.3 V so that they are compatible with the Raspberry Pi inputs. The problem here is that you also have to increase the Raspberry Pi SI, CS and MOSI output voltages from 3.3 V to 5 V since the MCP2525 requires at least 0.7 × 5 = 3.5 V for its high input voltage. This requires the use of voltage-level translator modules.

2. The other solution is to remove the TJA1050 chip from the interface module and replace it with a CAN bus transceiver module that can operate with 3.3 V (e.g., SN65HVD230). This requires some good electronic soldering experience!

3. The other solution, which is the one used in this book, is to separate the common +5 V power supply tracks on the interface module. The interface module is cheap and this is justified. Here, you have to cut the track at the back of the interface board that connects the MCP2515 chip power supply pin to TJA1050

power supply pin (Figure 9.3). The PCB track cutting is the responsibility of the reader! Then, solder a wire at the back of the board to the power supply pin of the TJA1050 chip (Figure 9.3). Alternatively, after cutting the track, you can connect a thin wire to one end of component C5 near pin 3 (power supply pin) of TJA1050 (Figure 9.3). You should then connect the VCC header pin to +3.3 V of Raspberry Pi (Pin 1), and the newly soldered wire to +5 V of Raspberry Pi (Pin 2). The newly soldered wire is shown as **NS*** in Figure 9.4.

Note: TJA1050 chip can draw up to 75 mA current. During the reboot, Raspberry Pi 4 can draw high currents from the power supply. If you have additional devices connected to your Raspberry Pi 4, such as an SSD drive, CPU fan and other interface devices, then the current provided by the power supply may not be enough and your Raspberry Pi 4 may not boot correctly. If this happens you should remove the TJA1050 +5 V pin from +5 V of the Raspberry Pi 4 until the Raspberry Pi 4 reboots and then connect it to +5 V of Raspberry Pi 4.

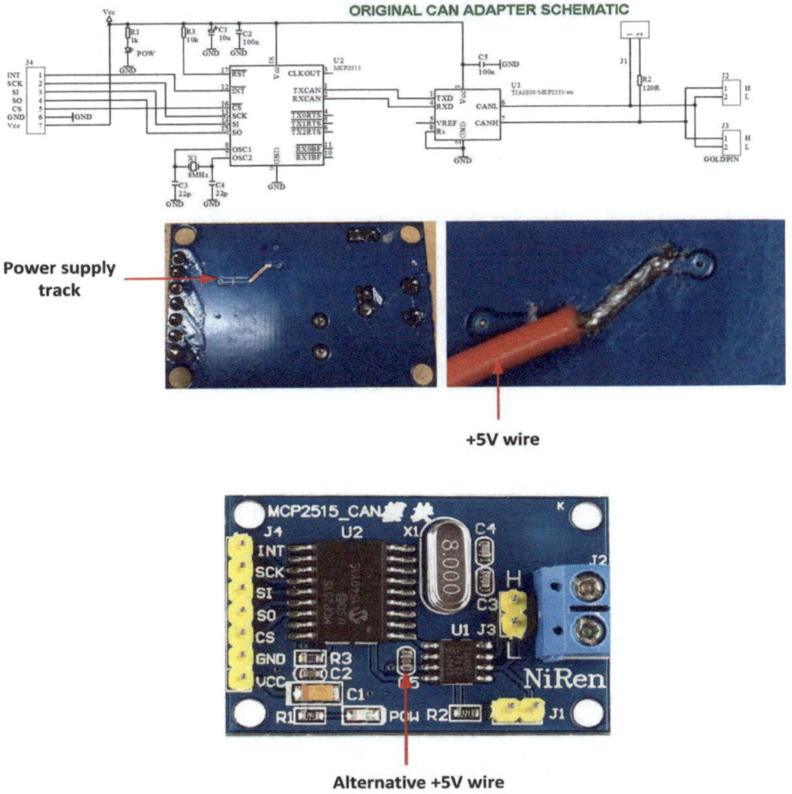

Figure 9.3: Separating the MCP2515 and TJA1050 power supply pins.

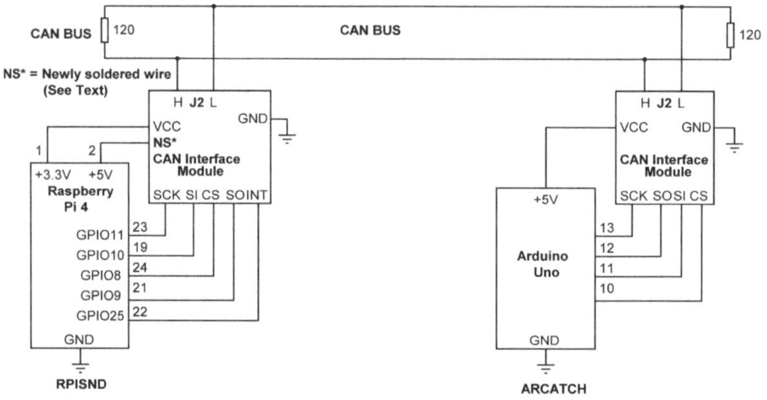

Figure 9.4: Circuit diagram of the project.

You have to enable the SPI bus and also install the MCP2515 library before you develop your Raspberry Pi program. The steps are:

- Enable the SPI bus by entering the command:

pi@raspberrypi:~ $ sudo raspi-config

- Scroll down to **3. Interface Options**.
- Select **SPI** and select **YES** to enable it.
- Select **Finish** to exit.
- Add the following line to file**: /boot/config.txt**

dtoverlay=mcp2515-can0,oscillator=8000000,interrupt=25

- Install the CAN utilities:

pi@raspberrypi:~ $ sudo apt-get install can-utils

CAN-utils includes applications such as:

- **candump** – Dump CAN packets (e.g., display, filter, and log to disk)
- **canplayer** – Replay CAN log files
- **cansend** – Send a single frame
- **cangen** – Generate random traffic
- **canbusload** – display the current CAN bus utilization

Connect the CAN bus interface module to your Raspberry Pi as shown in Figure 9.4 and enter the following command to set the bit rate to 250 KBPS. Common bitrates are: 10000, 20000, 50000, 100000, 125000, 250000, 500000, 800000, 1000000.

pi@raspberrypi:~ $ sudo /sbin/ip link set can0 up type can bitrate 250000

You should not get any error messages. The CAN interface behaves just like a network interface, and you should be able to get various statistics using the network configuration command **ifconfig** (see Figure 9.5)

pi@raspberrypi:~ $ ifconfig

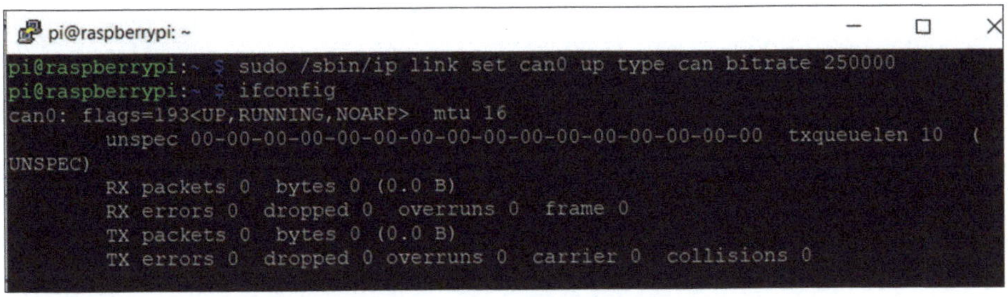

Figure 9.5: Command ***ifconfig****.*

To change a setting, first bring down the CAN interface by:

pi@raspberrypi:~ $ sudo ip link set can0 down

Node: ARCATCH Arduino Uno program listing

Figure 9.6 shows the program listing (Program: **ARCATCH**). Inside the **setup()** function, the CAN bus speed is set to 250 KBPS and Serial Monitor is initialized to 9600 baud and the text **Start** is displayed. Inside the main program loop, the program displays the data received over the CAN bus.

```
/*-----------------------------------------------------------
                      ARCATCH
                      ========

This program receives data over the CAN bus and displays the
message on the Serial Monitor

Author: Dogan Ibrahim
File  : ARCATCH
Date  : December 2022
-----------------------------------------------------------*/
#include <SPI.h>
#include <mcp2515.h>

#define CS 10
char msg;
struct can_frame MyMsg;
MCP2515 mcp2515(CS);                          // SPI CS
void setup()
{
```

```
  Serial.begin(9600);                          // Serial Monitor
  SPI.begin();
  mcp2515.reset();
  mcp2515.setBitrate(CAN_250KBPS, MCP_8MHZ);    // CAN config
  mcp2515.setNormalMode();                      // Normal mode
  Serial.println("Start");
}

void loop()
{
  if(mcp2515.readMessage(&MyMsg) == MCP2515::ERROR_OK)
  {
    Serial.println(«Received»);
    for(int j = 0; j < 8; j++)
    {
       msg = MyMsg.data[j];
       Serial.println(msg);
    }
  }
}
```

Figure 9.6: Program: ***ARCATCH.***

Node: RPISND: Command **cansend** is used in Raspberry Pi to transmit data over the CAN bus. The following command was entered on the terminal:

pi@raspberrypi:~ $ cansend can0 5a1#4142434445464748

Here, **can0** is the device name, **5a1** is the CAN ID, and the following bytes are the data bytes. Notice that the above command uses 11-bit ID. To use a 29-bit ID, change the ID to: **000005a1**

Testing the project: Figure 9.7 shows the data displayed on the Serial Monitor.

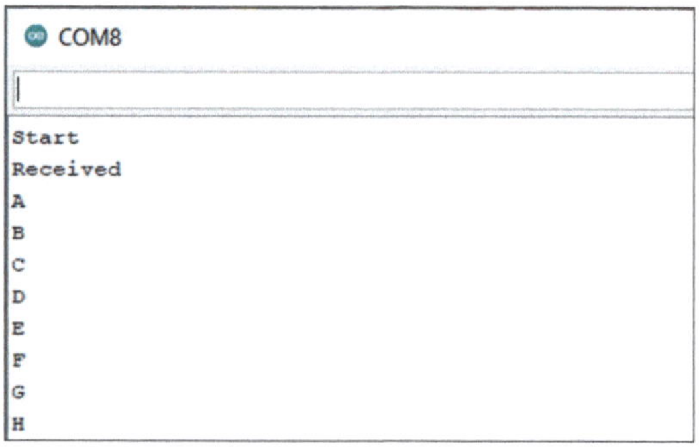

Figure 9.7: Displayed data.

Entering command **cansend** displays the format of this command as shown in Figure 9.8

```
pi@raspberrypi:~ $ cansend
cansend - send CAN-frames via CAN_RAW sockets.

Usage: cansend <device> <can_frame>.

<can_frame>:
 <can_id>#{data}          for 'classic' CAN 2.0 data frames
 <can_id>#R{len}          for 'classic' CAN 2.0 data frames
 <can_id>##<flags>{data}  for CAN FD frames

<can_id>:
 3 (SFF) or 8 (EFF) hex chars
{data}:
 0..8 (0..64 CAN FD) ASCII hex-values (optionally separated by '.')
{len}:
 an optional 0..8 value as RTR frames can contain a valid dlc field
<flags>:
 a single ASCII Hex value (0 .. F) which defines canfd_frame.flags

Examples:
  5A1#11.2233.44556677.88 / 123#DEADBEEF / 5AA# / 123##1 / 213##311223344 /
  1F334455#1122334455667788 / 123#R / 00000123#R3
```

Figure 9.8: Command: **cansend**.

9.3 Project 12: Raspberry Pi displaying messages from CAN bus

Description: In this project, two nodes are used: **ARSND** and **RPICATCH**. Node **ARSND** is the Arduino Uno node. This node transmits random numbers over the CAN bus. **ARCATCH** is the Raspberry Pi node that receives these numbers and displays them on the screen.

The block diagram and circuit diagram of this project are the same as Figure 9.1 and Figure 9.4, but nodes **RPISND** and **ARCATCH** are renamed as **ARSND** and **RPICATCH**, respectively.

Node: ARSND program listing

Figure 9.9 shows the Arduino Uno program. This program generates random numbers between 1 and 100 and transmits over the CAN bus.

```
/*----------------------------------------------------------
                    ARSND
                    =====

This program generates random numbers between 1 and 100 and
sends them over the CAN bus

Author: Dogan Ibrahim
File  : ARSND
Date  : December 2022
----------------------------------------------------------*/
#include <SPI.h>
#include <mcp2515.h>

#define CS 10
char RandomNumber;

struct can_frame MyMsg;
```

```
MCP2515 mcp2515(CS);                               // SPI CS

void setup()
{
  SPI.begin();
  mcp2515.reset();
  mcp2515.setBitrate(CAN_250KBPS, MCP_8MHZ);     // CAN bus config
  mcp2515.setNormalMode();                         // Normal mode
}

void loop()
{
  RandomNumber = random(1, 101);                   // Random number
  MyMsg.can_id = 0x30;                             // CAN ID
  MyMsg.can_dlc = 1;                               // 1 byte
  MyMsg.data[0] = RandomNumber;                    // Number in data[0]
  mcp2515.sendMessage(&MyMsg);                     // Send message
  delay(2000);                                     // Wait 2 secs
}
```

Figure 9.9: Program: ***ARSND.***

The following command was used on Raspberry Pi to read and display the numbers (make sure you set the bitrate before giving the following command):

pi@raspberrypi:~ $ candump can0

As shown in Figure 9.10, the program displays the device name (can0), CAN bus ID, number of bytes received, and the data received. Command **candump** has many options which can be listed by just entering **candump** on the terminal (Figure 9.11).

```
pi@raspberrypi:~ $ candump can0
  can0  030   [1]  43
  can0  030   [1]  11
  can0  030   [1]  26
  can0  030   [1]  27
  can0  030   [1]  2D
  can0  030   [1]  02
  can0  030   [1]  62
  can0  030   [1]  48
  can0  030   [1]  1D
  can0  030   [1]  26
  can0  030   [1]  3B
  can0  030   [1]  4E
```

Figure 9.10: Data received over the CAN bus.

```
pi@raspberrypi:~ $ candump
candump - dump CAN bus traffic.

Usage: candump [options] <CAN interface>+
  (use CTRL-C to terminate candump)

Options:
         -t <type>   (timestamp: (a)bsolute/(d)elta/(z)ero/(A)bsolute w date)
         -H          (read hardware timestamps instead of system timestamps)
         -c          (increment color mode level)
         -i          (binary output - may exceed 80 chars/line)
         -a          (enable additional ASCII output)
         -S          (swap byte order in printed CAN data[] - marked with '`' )
         -s <level>  (silent mode - 0: off (default) 1: animation 2: silent)
         -l          (log CAN-frames into file. Sets '-s 2' by default)
         -L          (use log file format on stdout)
         -n <count>  (terminate after reception of <count> CAN frames)
         -r <size>   (set socket receive buffer to <size>)
         -D          (Don't exit if a "detected" can device goes down.
         -d          (monitor dropped CAN frames)
         -e          (dump CAN error frames in human-readable format)
         -x          (print extra message infos, rx/tx brs esi)
         -T <msecs>  (terminate after <msecs> without any reception)
```

Figure 9.11: ***candump*** *options.*

You can display the CAN bus statistics by entering the command **ifconfig** (Figure 9.12).

```
pi@raspberrypi:~ $ ifconfig
can0: flags=193<UP,RUNNING,NOARP>  mtu 16
        unspec 00-00-00-00-00-00-00-00-00-00-00-00-00-00-00-00  txqueuelen 10  (
UNSPEC)
        RX packets 327  bytes 310 (310.0 B)
        RX errors 0  dropped 17  overruns 0  frame 0
        TX packets 1  bytes 4294967295 (3.9 GiB)
        TX errors 0  dropped 0 overruns 0  carrier 0  collisions 0
```

Figure 9.12: Displaying CAN bus statistics.

9.4 Project 13: Controlling LEDs connected to Arduino Uno

Description: In this project, there are two nodes: **ARLEDS** and **RPILED**. **ARLEDS** is the Arduino Uno node where two LEDs are connected to this node. **RPILED** sends commands over the CAN bus to turn ON these LEDs. An LED is turned ON for 3 seconds.

Block diagram: Figure 9.13 shows the block diagram of the project.

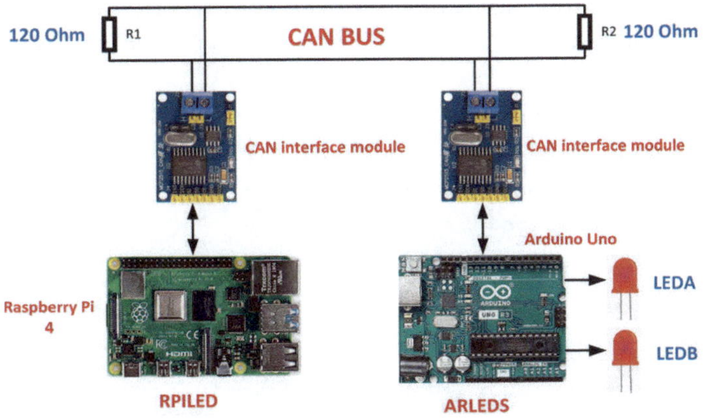

Figure 9.13: Block diagram of the project.

Circuit diagram: The circuit diagram is shown in Figure 9.14. **LEDA** is connected to port pin 2 and **LEDB** to port pin 3 of the Arduino Uno at node **ARLED**.

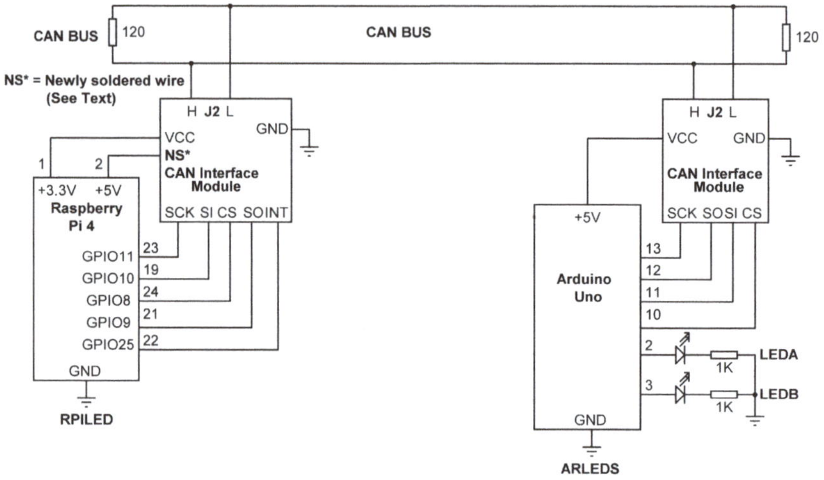

Figure 9.14: Circuit diagram of the project.

Node: RPILED commands (make sure you set the bitrate before giving the following commands):

To turn ON LEDA, enter the command: pi@raspberrypi:~ $ cansend can0 100#41

To turn ON LEDB, enter the command: pi@raspberrypi:~ $ cansend can0 100#42

Node ARLEDS program listing

Figure 9.15 shows the program listing (Program: **ARLEDS**). The CAN bus bitrate is set to 250 BPS. Inside the main program loop, the program checks if a message is available on the CAN bus. If the message is **A** then **LEDA** is turned ON for 3 seconds. If, on the other hand, the message is **B** then **LEDB** is turned ON for 3 seconds.

```
/*----------------------------------------------------------
                        ARLEDS
                        =====

This program receives messages over the CAN bus and
controls 2 LEDS at ports 2 and 3

Author: Dogan Ibrahim
File  : ARLEDS
Date  : December 2022
----------------------------------------------------------*/
#include <SPI.h>
#include <mcp2515.h>
```

```
#define CS 10
int LEDA = 2;
int LEDB = 3;
char resp;

struct can_frame MyMsg;
MCP2515 mcp2515(CS);                          // SPI CS

void setup()
{
  mcp2515.reset();
  mcp2515.setBitrate(CAN_250KBPS, MCP_8MHZ); // CAN bus config
  mcp2515.setNormalMode();                    // Normal mode
  pinMode(LEDA, OUTPUT);
  pinMode(LEDB, OUTPUT);
  digitalWrite(LEDA, 0);
  digitalWrite(LEDB, 0);
}

void loop()
{
  if(mcp2515.readMessage(&MyMsg) == MCP2515::ERROR_OK)
  {
    resp = MyMsg.data[0];
    if(resp == 'A')
    {
      digitalWrite(LEDA, 1);
      delay(3000);
      digitalWrite(LEDA, 0);
    }

    if(resp == 'B')
    {
      digitalWrite(LEDB, 1);
      delay(3000);
      digitalWrite(LEDB, 0);
    }
  }
}
```

Figure 9.15: Program: ***ARLEDS****.*

9.5 Using Python for CAN bus programs

In many applications, you may want to access the CAN bus by developing Python programs. Before doing this, you have to install the CAN bus Python library by entering the following command (there is an excellent document on the Internet called ***python-can Release 4.1.0*** which gives all the details of the Python CAN interface):

```
pi@raspberrypi:~ $ pip3 install python-can
```

The CAN bus library should then be imported using the command:

```
import can
```

An example is given below which transmits data bytes 1, 2, 3, 4, A over the CAN bus at 250 KBPS:

```
>>> import can
>>> bus=can.interface.Bus(bustype='socketcan',channel='can0',
    bitrate=250000)
>>> msg=can.Message(arbitration_id=100,data=[0x31,0x32,0x33,0x34,0x41],
    is_extended__id=False)
>>> bus.send(msg)
>>>
```

A receiver on CAN bus will get the following data bytes: **1234A**

A message can also be created using bytes or lists as shown below:

```
>>> tst = bytearray(b'Testing')
>>> msg=can.Message((arbitration_id=100,data=tst,is_extended__id=False)
>>> bus.send(msg)
>>>
```

The above statements will transmit bytes **Testing** over the CAN bus. You can get the message size by the statement:

```
>>> msg.dlc
7
```

To read a single message on the bus, use the statement:

```
>>> message = bus.recv()
```

The above statement is blocking and it will wait until message is available on the bus. For unblocking read use the statement:

```
>>> message = bus.recv(2.0)                 # 2 seconds timeout
```

If no message is received, you can display:

if message is None:
print(No message received, timeout")

9.6 Project 14: Controlling LEDs using buttons

Description: In this project, two nodes are used: **RPIBUTTONS** and **ARLEDS**. **RPIBUTTONS** is the Raspberry Pi node, which has got two pushbuttons labelled **BUTTONA** and **BUTTONB**. Pressing **BUTTONA** transmits character **A** over the CAN bus. Similarly, pressing **BUTTONB** transmits character **B** over the bus. Node **ARLEDS** is as in the previous project and turns ON **LEDA** for 3 seconds when **BUTTONA** is pressed, and **LEDB** when **BUTTONB** is pressed.

Block diagram: Figure 9.16 shows the block diagram of the project.

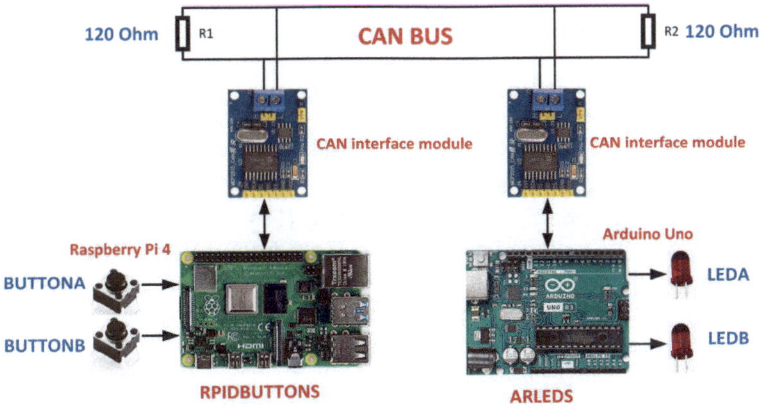

Figure 9.16: Block diagram of the project.

Circuit diagram: The circuit diagram is shown in Figure 9.17. LEDs are connected to Arduino Uno port pins 2 and 3. Buttons are connected to Raspberry Pi port pins GPIO2 and GPIO3 as shown in the figure.

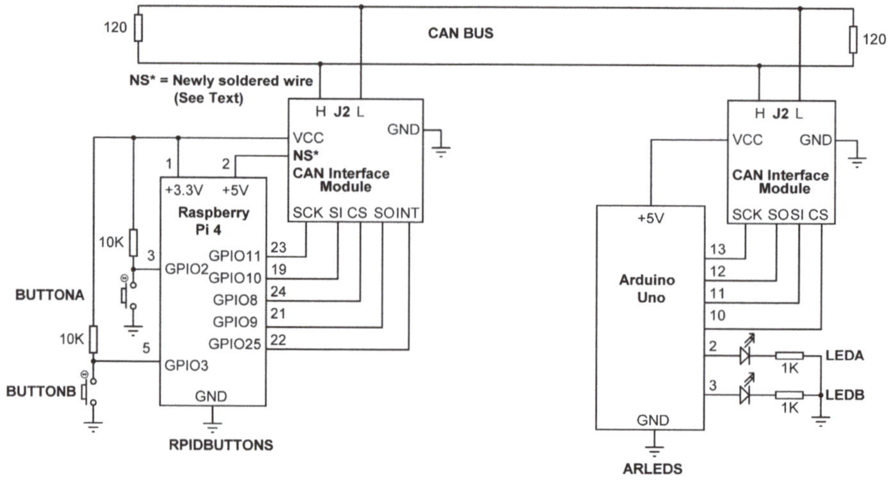

Figure 9.17: Circuit diagram of the project.

Node: ARLEDS program listing: The Arduino Uno program (Program: **ARLEDS**) is the same as the one in Figure 9.15.

Node: RPIBUTTONS program listing:
Figure 9.18 shows the Raspberry Pi program listing (Program: **RPIBUTTONS.py**). At the beginning of the program, modules **can** and **time** are imported into the program. **BUTTONA** and **BUTTONB** are configured as inputs. If **BUTTONA** is pressed, variable **stat** is set to **A**. Similarly, if **BUTTONB** is pressed, **stat** is set to **B**. The program then sends **stat** over the CAN bus. This process repeats until stopped by the user.

```
#==========================================================
#                    RPIBUTTONS
#                    ----------
#
# This program reads the state of two buttons named BUTTONA
# and BUTTONB. Pressing BUTTONA sends character A over the
# cab bus. Similarly, pressing BUTTONB sends character B over
# the CAN bus
#
# Author: Dogan Ibrahim
# File  : RPIBUTTONS.py
# Date  : December 2022
#============================================================
import RPi.GPIO as GPIO
GPIO.setwarnings(False)
GPIO.setmode(GPIO.BCM)

import can                                    # Import can
import time                                   # Import time
```

```
BUTTONA = 2                               # BUTTONA on port 2
BUTTONB = 3                               # BUTTONB on port 3

GPIO.setup(BUTTONA, GPIO.IN)              # Input
GPIO.setup(BUTTONB, GPIO.IN)              # Input

bus=can.interface.Bus(bustype='socketcan',channel='can0',bitrate=250000)

flag = 0
while True:
    if GPIO.input(BUTTONA) == 0:          # BUTTONA pressed?
        while GPIO.input(BUTTONA) == 0:   # BUTTONA released?
            pass
        stat = bytearray(b'A')            # To send A
        flag = 1

    if GPIO.input(BUTTONB) == 0:          # BUTTONB pressed?
        while GPIO.input(BUTTONB) == 0:   # BUTTONB released
            pass
        stat = bytearray(b'B')            # To send B
        flag = 1
#
# If a button is pressed then flag=1 and send msg to CAN bus
#
    if flag == 1:
        flag = 0
        msg = can.Message(arbitration_id=100, data=stat, is_extended_id=False)
        bus.send(msg)
        time.sleep(1)
```

Figure 9.18: Program: **RPIBUTTONS.py**.

Testing the project: Construct the circuit as shown in Figure 9.17 and apply power to both the Arduino Uno and Raspberry Pi. Make sure you set the bitrate to 250 BPS. Press the **BUTTONA** and **LEDA** should turn ON for 3 seconds. Then press **BUTTONB** and **LEDB** should turn ON for 3 seconds.

9.7 Project 15: CAN bus with 3 nodes: Controlling LEDs on Raspberry Pi node and Arduino Uno node

Description: In this project, there are 3 nodes: **ARDBUTTONS**, **RPILEDA**, and **ARDLEDB**. **ARDBUTTONS** and **ARDLEDB** are Arduino Uno-based nodes, and **RPILEDA** is a Raspberry Pi node. There are 2 buttons labelled BUTTONA and BUTTONB connected to node **ARDBUTTONS**. Pressing **BUTTONA** turns ON **LEDA** at Raspberry Pi node **RPILEDA**. Similarly, pressing **BUTTONB** turns ON **LEDB** at Arduino Uno node **ARDLEDB**.

Block diagram: Figure 9.19 shows the block diagram of the project.

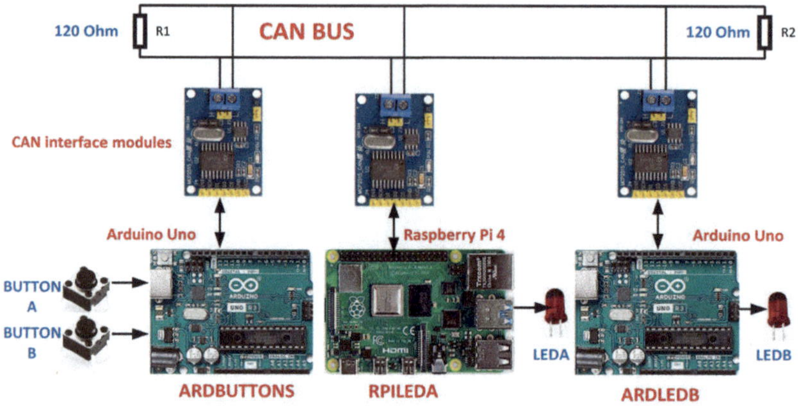

Figure 9.19: Block diagram of the project.

Circuit diagram: Figure 9.20 shows the circuit diagram of the project. **BUTTONA** and **BUTTONB** are connected to port pins 2 and 3 respectively of node **ARDBUTTONS**. **LEDA** is connected to port pin GPIO2 of Raspberry Pi. **LEDB** is connected to port pin 2 of node **ARDLEDB**.

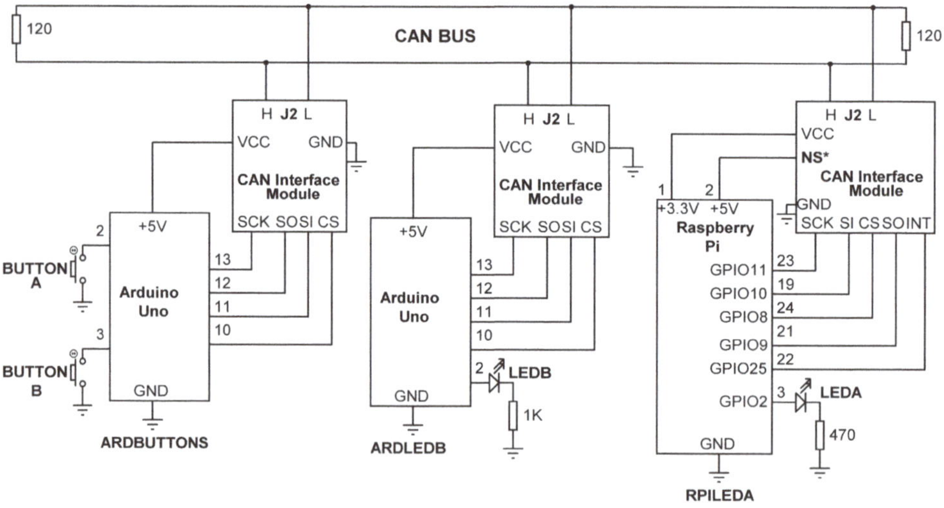

Figure 9.20: Circuit diagram.

Node: ARDBUTTONS program listing

Figure 9.21 shows the Arduino Uno program listing (Program: **ARDBUTTONS**). **BUTTONA** and **BUTTONB** are assigned to port pins 2 and 3, respectively, and are configured as inputs with the internal pull-up resistors enabled. The state of a button is therefore at logic HIGH when not pressed. Inside the **setup()** function, the CAN bus is configured to 250BPS. Inside the main program loop, the program checks if a button is pressed. If **BUTTONA** is pressed, then the message is set to 1 and the CAN ID is set to 0x50. If **BUTTONB** is pressed, then the message is set to 2 and the CAN ID is set to 0x60. The message is then transmitted over the CAN bus.

```
/*---------------------------------------------------------------
                        ARDBUTTONS
                        ==========

Two buttons are connected to this node. Pressing button BUTTO
sends a request over the CAN bus to turn ON LEDA. Similarly,
pressing BUTTONB sends a request to turn ON LEDB

Author: Dogan Ibrahim
File  : ARDBUTTONS
Date  : December 2022
-----------------------------------------------------------------*/
#include <SPI.h>
#include <mcp2515.h>

#define CS 10
int ReceivedMessage;
int BUTTONA = 2;                                   // R button
int BUTTONB = 3;                                   // G button
int T, ID, flag = 0;

struct can_frame MyMsg;
MCP2515 mcp2515(CS);

void setup()
{
  SPI.begin();
  mcp2515.reset();
  mcp2515.setBitrate(CAN_250KBPS, MCP_8MHZ);       // Config CAN bus
  mcp2515.setNormalMode();                         // Normal mode
  pinMode(BUTTONA, INPUT_PULLUP);                  // BUTTONA is input
  pinMode(BUTTONB, INPUT_PULLUP);                  // BUTTONB is input
}

void loop()
{
  if(digitalRead(BUTTONA) == 0)                    // BUTTONA pressed
  {
    while(digitalRead(BUTTONA) == 0);
    MyMsg.data[0] = 1;
    MyMsg.can_id = 0x50;                           // ID = 0x50
    flag = 1;
```

```
  }

  if(digitalRead(BUTTONB) == 0)                    // BUTTONB pressed
  {
    while(digitalRead(BUTTONB) == 0);
    MyMsg.data[0] = 2;
    MyMsg.can_id = 0x60;                           // ID = 0x60
    flag = 1;
  }

  if(flag == 1)                                    // If any button pressed
  {
    flag = 0;
    MyMsg.can_dlc = 1;                             // 1 bytes
    mcp2515.sendMessage(&MyMsg);                   // Send message
  }
}
```

Figure 9.21: Program: ***ARDBUTTONS.***

Node: ARDLEDB program listing

Figure 9.22 shows the Arduino Uno program listing (Program: **ARDLEDB**). The LED is connected to port pin 2 and is configured as an output. When a message is received over the CAN bus, the CAN ID and the message are checked. If the CAN ID is 0x50 and the message is 1, then it is assumed that the message is for this node, The LED is then turned ON for 3 seconds.

```
/*-------------------------------------------------------------
                       ARDLEDB
                       =======

 This program receives messages over the CAN bus and
 controls the LED connected to port pin 2

 Author: Dogan Ibrahim
 File  : ARDLEDB
 Date  : December 2022
--------------------------------------------------------------*/
 #include <SPI.h>
 #include <mcp2515.h>

 #define CS 10
 int LED = 2;
 char resp;

 struct can_frame MyMsg;
```

```
MCP2515 mcp2515(CS);                          // SPI CS

void setup()
{
  mcp2515.reset();
  mcp2515.setBitrate(CAN_250KBPS, MCP_8MHZ);  // CAN bus config
  mcp2515.setNormalMode();                    // Normal mode
  pinMode(LED, OUTPUT);
  digitalWrite(LED, 0);
}

void loop()
{
  if(mcp2515.readMessage(&MyMsg) == MCP2515::ERROR_OK)
  {
    resp = MyMsg.data[0];
    if(MyMsg.can_id == 0x50 && resp == 1)
    {
      digitalWrite(LED, 1);
      delay(3000);
      digitalWrite(LED, 0);
    }
  }
}
```

Figure 9.22: Program: ***ARDLEDB****.*

Node RPILEDA program listing

Figure 9.23 shows the Raspberry Pi program listing (Program: **RPILEDA**). At the beginning of the program, **LEDA** is assigned to port GPIO2 and is configured as an output. CAN bus filter is then set to 0x60 with the mask set to 0xFF so that only messages from node 0x60 will be accepted by this node. When a message is accepted, it is checked to be 2 and if so, the LED is turned ON for 3 seconds.

```
#=========================================================
#                    RPILEDA
#                    -------
#
# This program gets data over the CAN bus and controls an
# LED connected to port GPIO2. The LED is turned ON 3 secs
#
# Author: Dogan Ibrahim
# File  : RPILEDA.py
# Date  : December 2022
#===========================================================
```

```
import RPi.GPIO as GPIO
GPIO.setwarnings(False)
GPIO.setmode(GPIO.BCM)

import can                                              # Import can
import time                                             # Import time

LEDA = 2                                                # LEDA on port 2
GPIO.setup(LEDA, GPIO.OUT)                              # LED is output
GPIO.output(LEDA, 0)                                    # LED OFF

bus=can.interface.Bus(bustype='socketcan',channel='can0',bitrate=250000)

#
# Set ID to 0x60 and filter to 0xFF
#
filters=[{"can_id":0x60,  "can_mask": 0xFF, "extended":False}]

while True:
      dat = bus.recv()                                  # Receive data
      if dat.data == bytearray(b'\x02'):                # Data = 2?
         GPIO.output(LEDA, 1)                           # LEDA ON
         time.sleep(3)                                  # 3 secs
         GPIO.output(LEDA, 0)                           # LEDA OFF
```

Figure 9.23: Program: **RPILEDA**.

Testing the project: Construct the circuit as shown in Figure 9.20 and apply power to two Arduino Unos and to Raspberry Pi. Make sure you set the bitrate to 250 BPS. Press **BUTTONA** and **LEDA** should turn ON at node **RPILEDA** for 3 seconds. Then press **BUTTONB** and **LEDB** should turn ON at node **ARDLEDB** for 3 seconds.

Appendix

Electronic Components Used in the Projects

The following electronic parts are required for building the projects described in this book.

- 2× LM35DZ temperature sensor chip
- 1× RGB LED
- 1× DHT11 temperature and humidity sensor
- 1× I2C LCD
- 1× active buzzer
- 3× pushbutton
- 3× 1 k-ohm resistor
- 2× 120 ohm resistor
- 3× Arduino UNO board
- 3× CAN bus interface module
- Male-female breadboard jumper wires
- CAN bus wire
- Short CAN bus interface wires

Additionally required:

- 3× Arduino Uno with USB cables
- 1× Raspberry Pi with power supply

Index